Audio Watermark

AF400274

Yiqing Lin • Waleed H. Abdulla

Audio Watermark

A Comprehensive Foundation Using MATLAB

Springer

Yiqing Lin
The University of Auckland
Auckland, New Zealand

Waleed H. Abdulla
The University of Auckland
Auckland, New Zealand

Additional material to this book can be downloaded from http://extras.springer.com

ISBN 978-3-319-36094-2   ISBN 978-3-319-07974-5 (eBook)
DOI 10.1007/978-3-319-07974-5
Springer Cham Heidelberg New York Dordrecht London

Preface

Audio watermarking is a technique providing a promising solution to copyrights protection for digital audio and multimedia products. Using this technique, hidden information called *watermark* containing copyrights information is imperceptibly embedded into the audio track of a host media. This watermark may be extracted later on from a suspected media to verify the authenticity. To function as an effective tool to enforce ownership rights, the audio watermarking scheme must satisfy the imperceptibility, robustness, security, data payload, and computational complexity requirements. Throughout this book we will be illustrating in a practical way the commonly used and novel approaches of audio watermarking for copyrights protection. We will also introduce our recently developed methods for objectively predicting the perceptual quality of the watermarked audio signals.

This book is directed towards students, researchers, engineers, multimedia practitioners, and academics who are interested in multimedia authentication and audio pirating control. The theoretical descriptions of the watermarking techniques are augmented by MATLAB implementations to ease understanding of the watermarking principles. A GUI demonstration program for watermarking embedding and extraction under different attacks is also provided to quickly surf through the different aspects of the watermarking attributes.

Book Motivations and Objectives

Motivated by the booming of the digital media applications, plenty of research has been conducted to investigate the methods of audio watermarking for copyrights protection. However, clear and easy to follow information about the audio watermarking subject are still not widely available and scattered among many publications. Currently, it is hard to find an easy pathway to develop research in this field. One main reason to this difficulty is that most of the works are bounded by IP or patent constraints. On the implementation side it is still hard to find or write the implementation programs for the known audio watermarking techniques to see how the algorithms work. This book is introduced to establish a shortcut to get into this interesting field with minimal efforts. The commonly known techniques are well explained and supplemented with MATLAB codes to get a clear idea about how each technique performs. In addition, the reader can reproduce the functional figures of the book with provided MATLAB scripts written specifically for this purpose.

From the robustness and security perspectives, the commonly used audio watermarking techniques have limitations on the resistance to various attacks (especially desynchronization attacks) and/or security against unauthorized detection. Thus, in this book we develop new robust and secure audio watermark algorithm; it is well explained and implemented in MATLAB environment. This algorithm can embed unperceivable, robust, blind, and secure watermarks into digital audio files for the purpose of copyrights protection. In the developed algorithm, additional requirements such as data payload and computational complexity are also taken into account and detailed.

Apart from the improvement of audio watermarking algorithms, another landmark of this book is the exploration of benchmarking approaches to evaluate different algorithms in a fair and objective manner. For the application in copyrights protection, audio watermarking schemes are mainly evaluated in terms of imperceptibility, robustness, and security. In particular, the extent of imperceptibility is graded by perceptual quality assessment, which mostly involves a laborious process of subjective judgment. To facilitate the implementation of automatic perceptual measurement, we explore a new method for reliably predicting the perceptual quality of the watermarked audio signals. A comprehensive evaluation technique is illustrated to let the readers know how to pinpoint the strengths and weaknesses of each technique. The evaluation techniques are supported with tested MATLAB codes.

Furthermore to what we have just stated that this book extensively illustrates several commonly used audio watermarking algorithms for copyrights protection along with the improvement of benchmarking approaches, we may pinpoint the following new contributions of the current book:

- We introduce a spread spectrum based audio watermarking algorithm for copyrights protection, which involves Psychoacoustic Model 1, multiple scrambling, adaptive synchronization, frequency alignment, and coded-image watermark. In comparison with other existing audio watermarking schemes [1–10], the proposed scheme achieves a better compromise between imperceptibility, robustness, and data payload.
- We design a performance evaluation which consists of perceptual quality assessment, robustness test, security analysis, estimations of data payload, and computational complexity. The presented performance evaluation can serve as one comprehensive benchmarking of audio watermarking algorithms.
- We portray objective quality measures adopted in speech processing for perceptual quality evaluation of audio watermarking. Compared to traditional perception modelling, objective quality measures provide a faster and more

efficient method of evaluating the watermarked audio signals relative to host audio signals.

- We analyze methods for implementing psychoacoustic models in the MPEG standard, with the goal of achieving inaudible watermarks at a lower computational cost. With the same level of minimum masking threshold, Psychoacoustic Model 1 requires less computation time than Psychoacoustic Model 2.
- We identify the imperceptibility, robustness, and security characteristics of audio watermarking algorithms and further use them as attacks in the process of multiple watermarking.
- We propose the use of variable frame length to make the investigated cepstrum domain watermarking, wavelet domain watermarking, and echo hiding robust against time-scale modification.

Organization of the Book

The chapters in this book are organized as follows.

Chapter 1 provides an overview of digital watermarking technology and then opens a discussion on audio watermarking for copyrights protection.

Chapter 2 describes the principles of psychoacoustics, including the anatomy of the auditory system, perception of sound, and the phenomenon of auditory masking. Then two psychoacoustic models in the MPEG-1 standard, i.e., Psychoacoustic Model 1 and 2, are investigated. Through comparisons of the masking effect and the computational cost, the minimum masking threshold from Psychoacoustic Model 1 is chosen to be used for amplitude shaping of the watermark signal in Chap. 4.

Chapter 3 begins with the implementation specifications for perceptual quality assessment and the basic robustness test used in this chapter. Then it describes and evaluates several algorithms for audio watermarking, such as least significant bit modification, phase coding, spread spectrum watermarking, cepstrum domain watermarking, wavelet domain watermarking, echo hiding, and histogram-based watermarking. In the meantime, possible enhancements are exploited to improve the capabilities of some algorithms.

Chapter 4 presents a spread spectrum based audio watermarking algorithm for copyrights protection, which uses Psychoacoustic Model 1, multiple scrambling, adaptive synchronization, frequency alignment, and coded-image watermark. The basic idea is to embed the watermark by amplitude modulation on the time–frequency domain of the host audio signal and then detect the watermark by normalized correlation between the watermarked signal and corresponding secret keys.

In Chap. 5, the performance of the proposed audio watermarking algorithm is evaluated in terms of imperceptibility, robustness, security, data payload, and computational complexity. The evaluation starts with perceptual quality assessment, which consists of the subjective listening test (including the MUSHRA test and SDG rating) and the objective evaluation test (including the ODG by PEAQ and the SNR value). Then, the basic robustness test and the advanced robustness test (including a test with StirMark for Audio, a test under collusion, and a test under multiple watermarking) are carried out. In addition, a security analysis is followed by estimations of data payload and computational complexity. At the end of this chapter, a comparison between the proposed scheme and other reported systems is also presented.

Chapter 6 presents an investigation of objective quality measures for perceptual quality evaluation in the context of different audio watermarking techniques. The definitions of selected objective quality measures are described. In the experiments, two types of Pearson correlation analysis are conducted to evaluate the performance of these measures for predicting the perceptual quality of the watermarked audio signals.

Auckland, New Zealand
Auckland, New Zealand
Yiqing Lin
Waleed H. Abdulla

Contents

List of Figures

List of Tables

Chapter 1
Introduction

Since the last decade, online distribution of digital multimedia including images, audio, video, and documents has proliferated rapidly. In the open environment, it is easy to get free access to various information resources. Along with the convenience and high fidelity by which digital formatted data can be copied, edited, and transmitted, massive amounts of copyright infringements have arisen from illegal reproduction and unauthorized redistribution, which hinders the digital multimedia industry from progressing steadily [12]. To prevent these violations, the enforcement of ownership management has become an urgent necessity and is claiming more and more attention. As a result, digital watermarking has been proposed to identify the owner or distributor of digital data for the purpose of copyrights protection.

This chapter serves as an overall introduction to the book. First of all, background knowledge on information hiding, focusing on the differences between steganography and watermarking, is presented to ascertain the essence of watermarking. Then an overview of digital watermarking technology, including system framework, classifications, and applications, is introduced. Afterward, we focus on the requirements and benchmarking of audio watermarking for copyrights protection.

1.1   Information Hiding: Steganography and Watermarking

Information hiding is a general concept of hiding data in content. The term "hiding" can be interpreted as either keeping the existence of the information secret or making the information imperceptible [13]. Steganography and watermarking are two important subdisciplines of information hiding. Steganography seeks for ways to make communication invisible by hiding secrets in a cover message, whereas watermarking originates from the need for the copyrights protection of the content [14].

The word *steganography* is derived from the Greek *steganos* & *graphia*, which literally mean "covered writing." As defined in [13], steganography refers to the practice of undetectably altering a cover to embed a secret message, i.e., conveying hidden information in such a manner that nobody apart from the sender and intended

Y. Lin and W.H. Abdulla, *Audio Watermark: A Comprehensive Foundation Using MATLAB*, 1
DOI 10.1007/978-3-319-07974-5_1, © Springer International Publishing Switzerland 2015

recipient suspects the very existence of the message. Steganography has been used in a number of ways throughout time, for example, hidden tattoos, invisible inks, microdots, character arrangement, null ciphers, code words, covert channels, and spread spectrum communication [15, 16].

Note that steganography appears to be akin to cryptography, but not synonymous. Both cryptography and steganography are means to provide secrecy, but their methods of concealment are different. In cryptography, the message is encrypted to protect its content. One can tell that a message has been encrypted, but cannot decrypt it without the proper cipher. Once the data are decrypted, the protection is removed and there is no privacy any longer. In steganography, the message exists, but its presence is unknown to the receiver and others, such as the adversary. It is due to this lack of attention that the secret is well preserved. As stated in [15], "A cryptographic message can be intercepted by an eavesdropper, however, the eavesdropper may not even know a steganographic message exists." Therefore, steganography not only protects confidential information, as does cryptography, but also keeps the communicating parties safe to some extent. In the meantime, steganography and cryptography can be combined to provide two levels of security. That is, we encrypt a message using cryptography and then hide the encryption within the cover using steganography. This notion can be adopted in digital watermarking system to increase security.

Watermarking refers to the practice of imperceptibly altering an object to embed a message about that object [13], i.e., hiding specific information about the object without noticeable perceptual distortion. Watermarking has a long history dating back to the late thirteenth century, when "watermarks" were invented by paper mills in Italy to indicate the paper brand or paper maker and also served as the basis of authenticating paper. By the eighteenth century, watermarks began to be used as anticounterfeiting measures on money and other documents. So far, the most common form of paper watermark remains the bill in many countries. The first example of a technology similar to our notion of watermarks—imperceptible information about the objects in which they are embedded—was a patent filed for "watermarking" musical works by Emil Hembrooke in 1954. He inserted Morse code to identify the ownership of music, so that any forgery could be discerned. The term "digital watermarking" is the outcome of the digital era, which appears to have been first used by Komatsu and Tominaga in 1988. Since 1995, digital watermarking has gained a lot of attention and has evolved very fast [13, 14].

Watermarking and steganography are two areas of information hiding with different emphases. Both of them are required to be robust to protect the secret message. However, secrecy in watermarking is not strictly necessary, whereas steganography has to be secret by definition. For instance, it is preferred that everybody knows the presence of the watermark on bills and can recognize it easily against the light. Steganography requires only limited robustness as it generally relates to covert point-to-point communication between trusting parties, while watermarking must be quite robust to resist any attempts at removing the secret data as it is open to the public. Furthermore, the concealed message in watermarking is related to the object which has the same importance as itself. Therefore, no deterioration of the perceptual quality of the object is desired. But this is not compulsory in steganography, because the object there may be merely a carrier and has no intrinsic value [13, 14].

In the next section, we focus on digital watermarking, that is, watermarking applied to digital data. Key aspects will be discussed towards a deeper understanding.

1.2   Overview of Digital Watermarking

Digitization over all fields of technology has greatly broadened the notion of watermarking, and many new possibilities have been opened up. In particular, it is possible to hide information within digital image, audio, and video files in an unperceived and statistically undetectable sense. Driven by concerns over digital rights management (DRM), a new technique called digital watermarking has been put forward for intellectual property and copyrights protection [17, 18]. Digital watermarking is not designed to reveal the exact relationship between copyrighted content and the users, unless one violates its legal use.

Digital watermarking is the process of imperceptibly embedding watermark(s) into digital media as permanent signs and then extracting the watermark(s) from the suspected media to assure the authenticity [19]. The watermark(s) is always associated with the digital media to be protected or to its owner, which means that each digital media has its individual watermark or each owner has his/her sole watermark. For the purpose of copyrights protection, the advantage of digital watermarking over traditional steganography and cryptography is that digital media can be used in an overt manner, despite the presence of watermarks. In other words, we do not restrict the access to the watermarks residing in digital media, but make extra efforts to enhance their robustness against various attacks.

It is worth mentioning that some researchers provide another term closely related to the issue of copyrights protection, the so-called digital fingerprinting [20–23]. Fingerprints are characteristics of an object that tend to distinguish it from other similar objects. In a strict sense, fingerprinting refers to the process of identifying and recording fingerprints that are already intrinsic to the object[1] [14]. It is often regarded as a form of forensic watermarking used to trace authorized users who distribute them illicitly, i.e., the traitor tracing problem. Note that the greatest differences between digital watermarking and digital fingerprinting are the origin of hidden messages and operating mode. In digital watermarking, the watermark is an arbitrary message containing the information on proprietorship, while the fingerprint in digital fingerprinting is derived from the host itself and converted into a unique but much shorter number or string. Essentially, digital fingerprinting

---

[1]Although fingerprinting sometimes is related to the practice of extracting inherent features that uniquely identify the content, we avoid using this term to prevent confusion [13].

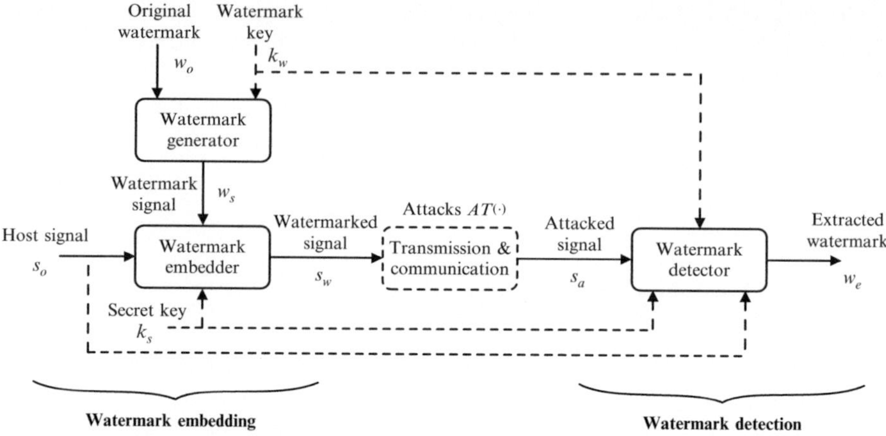

Fig. 1.1 A generic digital watermarking system

produces a metafile which describes the contents of the source file, so that a piece of work can be easily found and compared against other works in the database [17]. For this reason, digital fingerprinting was initially conceived for use in high-speed searching. Somewhat differently, digital watermarking stemmed from the motivation for copyrights protection of digital multimedia products. It is able to stand alone as an effective tool for copyright enforcement.

1.2.1   Framework of the Digital Watermarking System

In general, a digital watermarking system consists of three fundamental parts, namely a watermark generator, an embedder, and a detector, as illustrated in Fig. 1.1. Note that for different digital watermarking systems, the inputs indicated by dashed lines are optional.[2]

As a rule, the digital media to be protected is called a host signal, s_o, in which we choose to embed the original watermark, w_o. The form of the original watermark is diverse; possibly an image, a sequence of letters, or a simple series of bits. w_o can be rearranged into a collection of bits and further processed by a watermark generator in accordance with the watermark key, k_w, so as to generate the watermark signal, w_s. Usually, k_w is used as a kind of cryptography to offer additional protection. Then, the watermark embedder incorporates the watermark signal into the host

---

[2]Hereafter, if the items in the graphs are indicated by dashed lines, it means that they are optional.

Table 1.1 Classifications of digital watermarking

Basis for classification	Category
Type of medium to be watermarked	Image/ audio/ video/ text
Perceptibility	Imperceptible/ perceptible
Robustness of the watermark	Robust/ semi-fragile/ fragile
Need of host signal in the detection	Blind (public)/ non-blind (private)
Reversibility	Nonreversible/ reversible

signal, where the secret key k_s is employed to provide extra security and outputs the watermarked signal s_w. The embedding process is mathematically described as follows:

$$s_w = Embedding\ (s_o,\ w_o,\ k_w,\ k_s) \tag{1.1}$$

where s_w should be perceptually similar to s_o.

After that, the watermarked signal is spread out for communication. During the course of transmission, s_w is likely to be modified in some way, either being processed by common signal codings or tampered with by malicious attempts to remove the watermark. Such modifications are known collectively as "Attacks," $AT\ (\cdot)$, for instance, noise addition, MP3 compression, and random samples cropping.

In the detection, the watermark detector extracts the watermark from the signal received. The input to the watermark detector is called the attacked signal, s_a. The name is a general term, and s_a could be an identical or distorted version of s_w. The detection process is defined by

$$w_e = Detection\ (s_a,\ s_o,\ k_w,\ k_s) \tag{1.2}$$

where w_e is the extracted watermark. By comparing w_e with w_o, it is verified whether the host signal has been watermarked or not.

1.2.2 Classifications of Digital Watermarking

In terms of different characteristics, digital watermarking can be classified into several categories as summarized in Table 1.1[24–27].

- Image, audio, video, or text watermarking

There are different kinds of digital media that can be watermarked, such as image, audio, video, and text document. Image watermarking has developed well since the beginning of watermarking research. With relation to image watermarking, most current techniques for video watermarking treat video frames as a sequence of still images and watermark each of them accordingly. Compared to image and video watermarking, audio watermarking presents a special challenge due to less redundancy in audio files and the high sensitivity of the human auditory system (HAS). With the rapid development of audio compression techniques, audio products are becoming ever more popular on the Internet. Therefore, audio watermarking has attracted more and more attention in recent years. Text document watermarking also has applications wherever copyrighted electronic documents are distributed [17, 24, 28].

- Imperceptible or perceptible

For images and video, perceptible watermarks are visual patterns such as the logos merged into one corner of the images, ocular but not obstructive. Although perceptible watermarking is easy for practical implementation, it is not the focus of digital watermarking. As defined before, digital watermarking intends to imperceptibly embed the watermark into digital media [25].

- Robust, semi-fragile, or fragile

Watermark robustness accounts for the capability of the watermark to survive various manipulations. A robust watermark is a watermark that is hard to remove without deterioration of the original digital media. It is usually involved in copyrights protection, ownership verification, or other security-oriented applications. Conversely, a fragile watermark is a watermark that is vulnerable to any modification, mainly for the purpose of data authentication. In a temperate manner, a semi-fragile watermark is marginally robust and moderately sensitive to some attacks [24–26].

- Blind (public) or non-blind (private)

Blind (public) digital watermarking does not require the host signal for watermark detection. On the contrary, digital watermarking that requires the host signal to extract the watermark is non-blind (private). Generally, watermark detection is more robust if the original unwatermarked data are available. However, access to the original host signal can not be warranted in most real-world scenarios. Therefore, blind watermarking is more flexible and practical [24, 28].

- Nonreversible or reversible

In reversible watermarking, the watermark can be completely removed from the watermarked signal, thus allowing it to obtain an exact recovery of the host signal. However, the price of such reversibility implicates some loss of robustness and security. Nonreversible watermarking usually introduces a slight but irreversible degradation in the original signal. Watermark reversibility must only be considered in applications where complete restoration of the host signal is in great request [24, 27].

1.2.3 Applications of Digital Watermarking

Digital watermarking can be used in a wide range of applications. There is no denying that other techniques might be viable alternatives sometimes. However, the attributes of digital watermarking make it indispensable for certain purposes.

1.2.3.1 Copyrights Protection

The exploration of digital watermarking was driven by the desire for copyrights protection. The idea is to embed a watermark with copyright information into the media. When proprietorial disputes happen, the watermark can be extracted as reliable proof to make an assertion about the ownership. To this end, the watermark must be inseparable from the host and robust against various attacks intended to destroy it. Moreover, the system requires a high level of security to survive the statistical detection. With these properties, the owner could demonstrate the presence of watermark to claim the copyright on the disputed media. In addition, since it is not necessary for the watermark to be very long, the data payload for this application does not have to be high [29, 30].

1.2.3.2 Content Authentication

In authentication application, the objective is to verify whether the content has been tampered with or not. Since the watermarks undergo the same transformations as the host media, it is possible to learn something about the occurrences by looking at the resulting watermarks. For this purpose, fragile watermarks with a low robustness are commonly employed. If the content is manipulated in an illegal fashion, fragile watermarks will be changed to reveal that the content is not authentic [13, 14].

1.2.3.3 Broadcast Monitoring

The target of broadcast monitoring is to collect information about the content being broadcast. This information is then used as the evidence to verify whether the content was broadcast as agreed or for some other purposes, such as billing or statistical analysis for product improvement. In this case, the robustness of the watermark is not a concern due to a lower risk of distortion. Instead, transparent or unnoticeable watermarks, i.e., imperceptibility, are more required [13, 31].

1.2.3.4 Copy Control

Most applications of digital watermarking, as discussed so far, have an effect only after the infringement has happened. In the copy control application, the aim is to prevent people from making illegal copies of copyrighted media. The mechanism is to embed watermarks indicating copy status of the content in copyright compliant devices, proposed by the Secure Digital Music Initiative (SDMI). For example, if the DVD system contains the data with copyright information embedded as watermarks, then a compliant DVD player will not play back or copy data that carry a "copy never" watermark [13, 14, 30].

Digital watermarking also has been applied in device control, legacy enhancement, transaction tracking (or fingerprinting), and so on. More details can be found in [13, 14, 24, 27, 30].

1.3 Audio Watermarking for Copyrights Protection

Compared to image and video watermarking, inserting watermark(s) into digital audio files is a more arduous task. Generally, the human auditory system is much more sensitive than the human visual system (HVS), implying that inaudibility is much more difficult to achieve than invisibility for images. Moreover, audio signals are represented by far less samples per time interval, and thereby the amount of information that can be embedded robustly and inaudibly is much lower than for visual media [28].

Audio watermarking is a promising solution to copyrights protection for digital audio and multimedia products. To function as an effective tool to enforce ownership rights, any eligible audio watermarking scheme must meet a number of requirements to be described in Sect. 1.3.1. The benchmarking of any audio watermarking technique is measured against these requirements.

1.3.1 Requirements for the Audio Watermarking System

The audio watermarking system for copyrights protection has to comply with the following main requirements: excellent imperceptibility for preserving the perceptual quality of the audio file, strong robustness against various attacks, and high-level security for preventing unauthorized detection. Data payload and computational complexity are two additional criteria [30].

1.3.1.1   Imperceptibility

Imperceptibility is a prerequisite to practicality. The process of audio watermarking is considered to be imperceptible or transparent if no differences between the host and watermarked signals are perceivable. Otherwise it is perceptible or nontransparent. To preserve the perceptual quality of the watermarked data, a psychoacoustic model derived from the auditory masking phenomenon will be relied on to deceive the human perception of digital audio files [32]. Consequently it appears as if there is nothing added to the host media.

1.3.1.2   Robustness

Robustness is a measure of reliability and refers to the capability of resisting a variety of unintentional and intentional attacks. In other words, the watermark detector should be able to extract the watermark from the attacked watermarked signal. Examples of attacks on audio watermarking include many kinds of signal processing and coding, such as noise addition, resampling, requantization, MPEG (Moving Picture Experts Group) compression, random samples cropping[3], time-scale modification (TSM), and pitch-scale modification (PSM). The last three attacks belong to desynchronization attacks, which introduce displacement and heavily threaten the survival of the watermark.

1.3.1.3   Security

Security is a prerequisite to existence. Since the watermarking algorithms are likely to be open to the public, we should guarantee that the watermarks cannot be ascertained even by reversing the embedding process or performing statistical detection [33,34]. In this case, secret keys (usually pseudorandom sequences) and/or scrambling operations can be adopted to add randomness into the embedding and detection processes, so that the digital watermarking system is self-secured.

1.3.1.4   Data Payload

Data payload refers to the amount of bits carried within a unit of time [13]. In digital audio watermarking, it is defined as the number of bits embedded in a one-second audio fraction, expressed in bit per second (bit/s or bps). Data payload of the audio watermarking system varies greatly, depending on the embedding parameters and

---

[3]Random samples cropping includes deliberate removal of the header or footer of a signal. Therefore, the watermark should be spread throughout the entire audio signal.

the embedding algorithm. Copyrights protection applications do not require a high data payload, only $2 \sim 4$ bits/s on average [35].

1.3.1.5 Computational Complexity

From a technological point of view, the computational complexity of a watermarking system involves two principal issues of consideration. One is the speed with which embedding and detection are performed, and the other is the number of embedders and detectors [13], where the speed is more of concern for us. The most intuitive way to estimate the speed is to separately measure embedding and detection time relative to the duration of the host audio. For a fair comparison, the measurements should be carried out on platforms with the same computational capabilities. Although a real-time and low-delay system is commonly desired, different applications require different speeds. For the purpose of copyrights protection, even a commercial product does not care too much about the embedding time. Conversely, the customers expect to extract the watermark as quickly as possible.

In practice, no one system can fully satisfy all the requirements and some trade-offs always exist among criteria. Typically, an audio watermarking system can operate with either excellent imperceptibility or strong robustness, but not both. In order to ensure the robustness, we embed the watermark(s) into perceptually important regions or increase the strength of the watermarking. However, such strategies are liable to cause perceivable distortion to the host signal, which is against the property of imperceptibility. Moreover, both of them are in close connection with data payload. If we embed more bits into an audio signal, the imperceptibility would become worse and the robustness would be stronger[36]. Similar compromises also occur between imperceptibility, robustness, and security.

1.3.2 Benchmarking on Audio Watermarking Techniques

Along with the advancement of audio watermarking techniques, the necessity for benchmarking various algorithms effectively and comprehensively becomes imperative [37, 38]. Since appropriate assessment criteria always depend on the application, it is impractical and inaccurate to develop a universal benchmark for all kinds of digital watermarking systems [13]. As discussed above, imperceptibility, robustness, and security are key principles in designing any audio watermarking scheme for the application of copyrights protection. Accordingly, performance evaluations in our research are focused on those three aspects.

Table 1.2 Subjective
difference grade (SDG)

Difference grade	Description of impairments
0	Imperceptible
−1	Perceptible but not annoying
−2	Slightly annoying
−3	Annoying
−4	Very annoying

1.3.2.1   Perceptual Quality Assessment

Similar to evaluating the quality of perceptual codecs in the audio, image, and
video fields [39], perceptual quality assessment on the watermarked audio files
is usually classified into two categories: subjective listening tests by human
acoustic perception and objective evaluation tests by perception modelling or quality
measures. Both of them are indispensable to the perceptual quality evaluation of
audio watermarking.

As perceptual quality is essentially decided by human opinion, subjective
listening tests on audiences from different backgrounds are required in most
applications [39]. In subjective listening tests, the subjects are asked to discern
the watermarked and host audio clips. Two popular modes are the ABX test
[40, 41] and the MUSHRA test (i.e., MUlti Stimuli with Hidden Reference and
Anchors) [42], derived from ITU-R Recommendation BS.1116 [43] and BS.1534
[44][4], respectively. Moreover, the watermarked signal is graded relative to the host
signal according to a five-grade impairment scale (see Table 1.2) defined in ITU-R
BS.562[5]. It is known as the subjective difference grade (SDG), which equals to the
subtraction between subjective ratings given separately to the watermarked and host
signals. Therefore, SDG near 0 means that the watermarked signal is perceptually
undistinguished from the host signal, whereas SDG near −4 represents a seriously
distorted version of the watermarked signal.

However, such audibility tests are not only costly and time-consuming, but also
heavily depend on the subjects and surrounding conditions [46]. Therefore, the
industry desires the use of objective evaluation tests to achieve automatic perceptual
measurement. Currently, the most commonly used objective evaluation is perception
modelling, i.e., assessing the perceptual quality of audio data via a stimulant ear,
such as Evaluation of Audio Quality (EAQUAL) [47], Perceptual Evaluation of
Audio Quality (PEAQ) [48], and Perceptual Model-Quality Assessment (PEMO-Q)
[49]. Moreover, objective quality measures are exploited as an alternative approach
to quantify the dissimilarities caused by audio watermarking. For instance, a widely
used quality measure is the signal-to-noise ratio (SNR), calculated as follows [50]:

---

[4]ITU-R: Radiocommunication Sector of the International Telecommunication Union; BS: Broadcasting service (sound).

[5]ITU-R BS.562 has been replaced by ITU-R BS.1284[45].

$$SNR\left(s_w, s_o\right) = 10 \cdot log_{10} \frac{\sum_n \left[s_o\left(n\right)\right]^2}{\sum_n \left[s_w\left(n\right) - s_o\left(n\right)\right]^2} \qquad (1.3)$$

where $\sum_n \left[s_o\left(n\right)\right]^2$ is the power of host signal s_o and $\sum_n \left[s_w\left(n\right) - s_o\left(n\right)\right]^2$ is the power of noise caused by watermarking.

1.3.2.2 Robustness Test

The goal of the robustness test is to test the ability of a watermarking system resistant to signal modifications in real applications. In the robustness test, various attacks are applied to the watermarked signal and produce a number of attacked signals. Then, watermark detection is performed on each attacked signal to check whether the embedded watermark survives or not. In particular, the detection rate is denoted by bit error rate (BER), defined in the following equation:

$$BER = \frac{\textit{Number of wrong bits between } w_e \textit{ and } w_o}{\textit{Number of bits of } w_o} \times 100\% \qquad (1.4)$$

A competent robustness test should comprise an extensive range of possible attacks. Tens of attacks are employed in some popular audio watermarking evaluation platforms, i.e. SDMI standard, STEP 2000 and StirMark for Audio, which are described in Appendix A, B and C respectively. In summary, typical signal manipulations on audio watermarking schemes are classified into three categories: common signal operations (such as noise addition, resampling, requantization, amplitude scaling, low-pass filtering, echo addition, reverberation, MP3 compression, DA/AD conversion, and combinations of two or more), desynchronization attacks (such as random samples cropping, jittering, zeros inserting, time-scale modification and pitch-scale modification), and advanced attacks (such as collusion and multiple watermarking[6]). In most cases, a robustness test on an audio watermarking system includes the first two kinds of attacks, while the last kind is only taken into consideration for some specific applications. Moreover, desynchronization attacks are more challenging for most audio watermarking systems. Loss of synchronization would cause mismatch in positions between watermark embedding and detection, which is disastrous to watermark retrieval [19].

It is worthy of notice that there is a premise for undertaking a robustness test. That is, the degree of deterioration by attacks should keep within an acceptable limit, because it is needless for detection to proceed on a watermarked signal that is already severely destroyed. Therefore, attack parameters should control the amplitude of noise added and the extent of stretching or shifting within certain limits.

---

[6]To be described in section 3.1.3.2, multiple watermarking is to embed several watermarks sequentially.

1.3.2.3   Security Analysis

Security analysis is performed to evaluate the characteristics of security for audio watermarking systems. Since security is attributed to the randomness merged by sequences of pseudorandom numbers (PRN) and/or scrambling operations, an intuitive method of security analysis is to calculate the number of possible embedding ways. If there are more possible ways of embedding, it would be difficult for unauthorized detection to ascertain the embedded watermark. This indicates that the system has a high level of security.

Note that in the performance evaluation, a variety of audio signals have to be involved to truly verify the properties of the audio watermarking system. The test set should be representative of a typical range of audio content [13], such as classical, rock and folk music, vocal and instrumental music, and so on.

Chapter 2
Principles of Psychoacoustics

Psychoacoustics is the science of sound perception, i.e., investigating the statistical relationships between acoustic stimuli and hearing sensations [51]. This study aims to build up the psychoacoustic model, a kind of quantitative model, which could closely match the hearing mechanism. A good understanding of the sensory response of the human auditory system (HAS) is essential to the development of psychoacoustic models for audio watermarking, where the perceptual quality of processed audio must be preserved to the greatest extent.

In this chapter, the basic structure and function of the auditory system, mainly the peripheral part, are illustrated for the comprehension of human hearing. Then, the hearing threshold and auditory masking phenomenon are analyzed to pave the way for deriving the psychoacoustic models. Finally, Psychoacoustic Model 1 in ISO/MPEG standard is implemented to be utilized in our audio watermarking scheme later on.

2.1 Physiology of the Auditory System

Hearing is the sense by which sound is perceived [52]. Human hearing is performed primarily by the auditory system, in which the peripheral part is of more relevance to our study. The peripheral auditory system (the ear, that portion of the auditory system not in the brain [53]) includes three components: the outer ear, the middle ear, and the inner ear, as illustrated in Fig. 2.1.

The whole process of capturing the sound through the ear to create neurological signals is an intricate and ingenious procedure. First, the sound wave travels through the auditory canal and causes the eardrum to vibrate. This vibration is transmitted via the ossicles of the middle ear to the oval window at the cochlea inlet. The movement of the oval window forces the fluid in the cochlea to flow, which results in the vibration of the basilar membrane that lies along the spiral cochlea. This motion causes the hair cells on the basilar membrane to be stimulated and to generate neural

Y. Lin and W.H. Abdulla, *Audio Watermark: A Comprehensive Foundation Using MATLAB*,   15
DOI 10.1007/978-3-319-07974-5_2, © Springer International Publishing Switzerland 2015

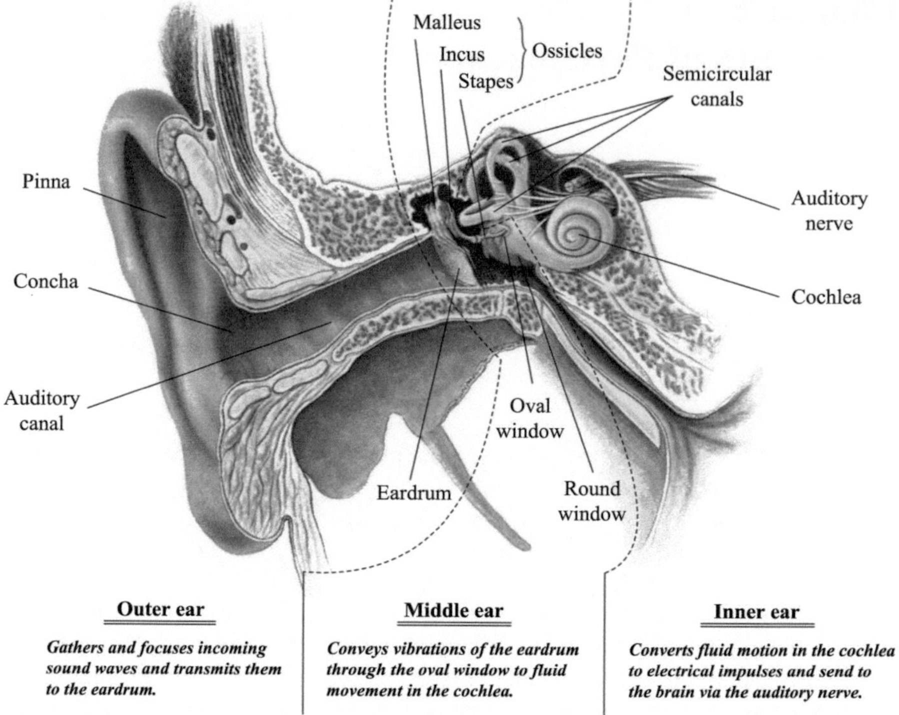

Fig. 2.1 Structure of the peripheral auditory system [57]

responses carrying the acoustic information. Then, the neural impulses are sent to the central auditory system through the auditory nerves to be interpreted by the brain [54, 55].

2.1.1 The Outer Ear

Sounds communicate the auditory system via the outer ear. The pinna and its deep center portion, the concha, constitute the externally visible part of the outer ear that serves focusing the sound waves at the entrance of the auditory canal (or auditory meatus). Since human pinna has no useful muscles, it is nearly immobile. Therefore, the head must be reoriented towards the direction of acoustical disturbance for a better collection and localization of sound. The auditory canal (usually 2–3 cm in length) is a tunnel through which the sound waves are conducted, and it is closed with the eardrum (or tympanic membrane).[1] The eardrum is stretched tightly across the inner end of the auditory canal and is pulled slightly inward by structures in the middle ear [58]. Upon travelling through the auditory canal, sound waves impinge on the eardrum and cause it to vibrate. Then, these mechanical vibrations which respond to the pressure fluctuations of acoustic stimuli are passed along to the middle ear.

The outer ear plays an important role in human hearing. The pinna is of great relevance to sound localization, since it reflects the arriving sound in ways that depend on the angle of the source. The resonances occurring in the concha and auditory canal bring about an increase on sound pressure level (SPL) for frequencies between 1.5 kHz and 7 kHz. The extent of amplification depends on both the frequency and angle of the incident wave, as indicated in Fig. 2.2. For example, the gain is about 10–15 dB in the frequency range from 1.5 kHz to 7 kHz at an azimuthal angle of 45°. Moreover, the outer ear protects the eardrum and the middle ear against extraneous bodies and changes in humidity and temperature [59].

2.1.2 The Middle Ear

The eardrum vibrations are transferred through the middle ear to the inner ear. The middle ear is an air-filled chamber, bounding by the eardrum laterally and by the oval window of the cochlea medially. It contains three tiny bones known as the ossicles: the malleus (or hammer), incus (or anvil), and stapes (or stirrup). These three ossicles are interconnected sequentially and suspended in the middle ear cavity by ligaments and muscles. As shown in Fig. 2.1, the malleus is fused to the eardrum and articulates with the incus; the incus is connected to both the other bones; the stapes is attached to the incus and its footplate fits into the oval window of the cochlea. The oval window is a membrane-covered opening which leads from the middle ear to the vestibule of inner ear.

As an interface between the outer and inner ears, the middle ear has two functions. One function is to serve as an impedance-matching transformer that ensures an efficient transmission of sound energy. As we know, the outer and middle ear cavities are filled with air, while the inner ear is filled with fluid. So the passage of pressure waves from the outer ear to the inner ear involves a boundary between air and fluid, two mediums with different acoustic impedance.[2] In fact, approximately 99.9 % of sound energy incident on air/fluid boundary is reflected back within the air medium, so that only 0.1 % of the energy is transmitted to the fluid. It means that

---

[1] In this sense, the auditory canal closed with the eardrum at its proximal end has a configuration as a resonator.

[2] Acoustic impedance is a constant related to the propagation of sound waves in an acoustic medium. Technically, sound waves encounter much less resistance when travelling in air than in fluid.

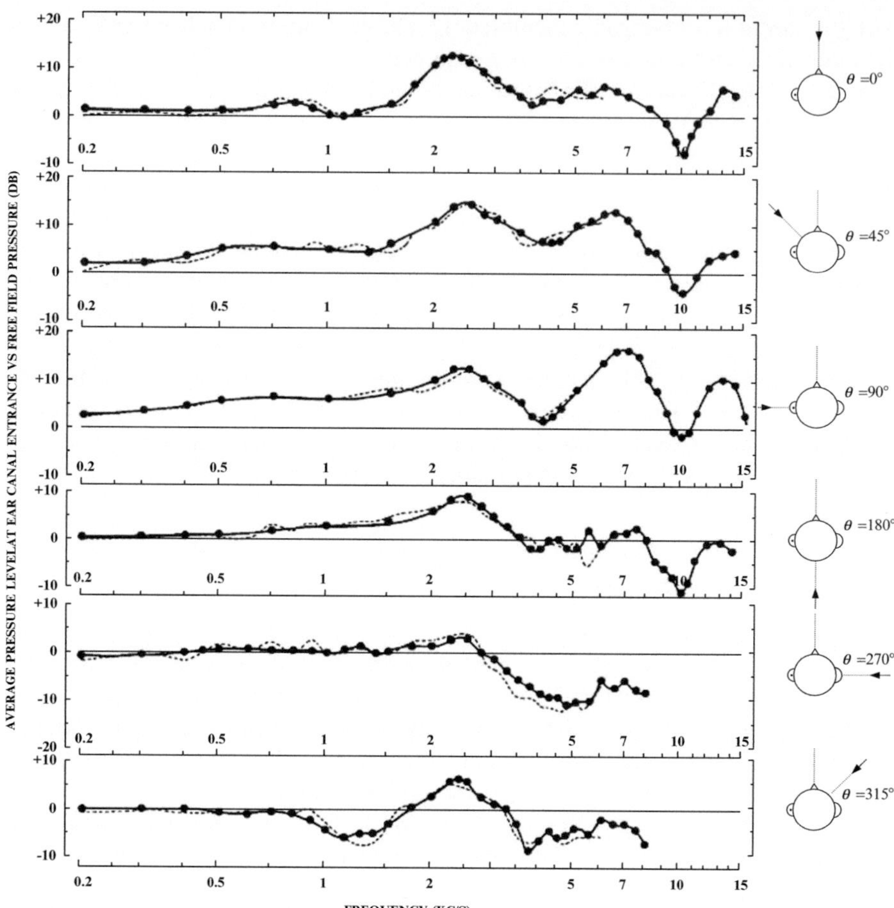

Fig. 2.2 Average pressure levels at auditory canal entrance versus free-field pressure, at six azimuthal angles of incidence [60]. *Notes*: (1) The sound pressure was measured with a probe tube located at the left ear of the subject. (2) A point source of sound was moved around a horizontal circle of radius 1 m with the subject's head at the center. At $\theta = 0°$, the subject was facing the source, and at $\theta = 90°$, the source was normally incident at plane of left ear

if sound waves were to hit the oval window directly, the energy would undergo a loss of 30 dB before entering the cochlea. To minimize this reduction, the middle ear has two features to match up the low impedance at the eardrum with high impedance at the oval window. The first is related to the relative sizes of the eardrum and the stapes footplate which clings to the oval window. The effective area of the eardrum is about 55 mm^2 and that of the footplate is about 3.2 mm^2; thereupon they differ in size by a factor of 17 (55 mm^2/3.2 mm^2 = 17). So, if all the force exerted on the eardrum is transferred to the footplate, then the pressure (force per unit area) at the oval window is 17 times greater than at the eardrum. The second depends on the lever action of the ossicular chain that amplifies the force of the incoming auditory signals. The lengths of the malleus and incus correspond to the distances from the pivot to the applied and resultant forces, respectively. Measurements indicate that the ossicles as a lever system increases the force at the eardrum by a factor of 1.3. Consequently, the combined effect of these actions effectively counteracts the reduction caused by the impedance mismatch [58]. Another function of the middle ear is to diminish the transmission of bone-conducted sound to the cochlea by muscle contraction. If these sounds were sent over to the cochlea, they would appear very loud that may be harmful to the inner ear [61].

2.1.3 The Inner Ear

The inner ear transduces the vibratory stimulation from the middle ear to neural impulses which are transmitted to the brain. The vestibular apparatus and the cochlea are the main parts in the inner ear. The vestibular apparatus is responsible for the sense of balance. It includes three semicircular canals and the vestibule. The cochlea is the central processor of the ear, where the organ of corti, the sensory organ of hearing, is located. The cochlea is a spiral-shaped bony tube structure of decreasing diameter, which coils up $2\frac{3}{4}$ times around a middle core containing the auditory nerve, as shown in Fig. 2.3a.[3] The duct is filled with almost incompressible fluids and is enclosed by the oval window (the opening to the middle ear) and the round window (a membrane at the rear of the cochlea). When the stapes pushes back and forth on the oval window, the motion of the oval window causes the fluid to flow and impels the round window to move reciprocally, which lead to the variations of fluid pressure in the cochlea. The movements of the oval and round windows are indicated by the solid and dotted arrows in Fig. 2.3a.

Figure 2.3c shows the cross-section through one cochlea turn. Two membranes, Reissner's membrane and the basilar membrane, divide the cochlea along the spiral direction into three fluid-filled compartments: scala vestibuli, scala media, and scala tympani. The scala vestibuli and scala tympani are merged through a small opening called helicotrema at the apex, and they contain the same fluid (the perilymph) with most of the nervous system. The scala media is segregated from other scalae and contains a different fluid (the endolymph). On the scala media surface of basilar membrane (BM) lies the organ of corti. The changes of fluid pressure in the cochlea will cause the BM to deform, so that the hair cells[4] on the organ of corti are

---

[3]Note that the cochlea is a cavity within the skull, not a structure by itself [58]. Hence the unraveled cochlea in Fig. 2.3b is impossible in practice, only for the sake of illustration.

[4]The hair cells including the outer and inner hair cells (OHC and IHC) are auditory receptors on the organ of corti.

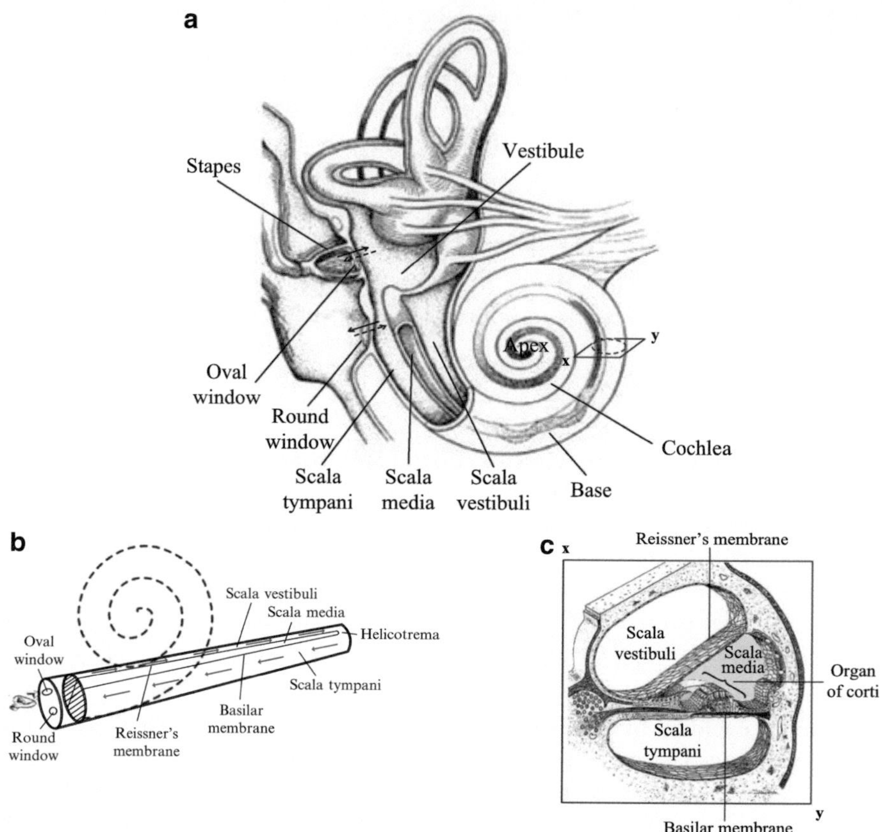

Fig. 2.3 Anatomy of the cochlea (**a**) Relative location of the cochlea in the inner ear [61] (**b**) Schematic of the unraveled cochlea (**c**) Cross-section through one cochlea turn [65]

stimulated to transduce the movement of the BM into neural impulses. Then the neural signals are carried over to the brain via auditory nerve, which ultimately lead to the perception of sound.

The basilar membrane extends along the spirals of the cochlea and is about 32 mm long. It is relatively narrower and stiffer at the base (near the windows), while it gets wider and more flexible at the apex (near the helicotrema). Accordingly, each location on the BM has different vibratory amplitude in response to sound of different frequencies, which means that each point resonates at a specific characteristic frequency (CF) [54]. As exemplified in Fig. 2.4a, for high-frequency tones, the maximum displacement of the BM occurs near the base, with tiny movement on the remainder of the membrane. For low-frequency tones, the vibration travels all the way along the BM, reaching its maximum close to the apex.[5] Figure 2.4b

---

[5]There is one fact worth of attention, i.e., any location on the BM will respond to a wide range of tones that are lower than its CF. That's why low frequencies are less selective than high frequencies.

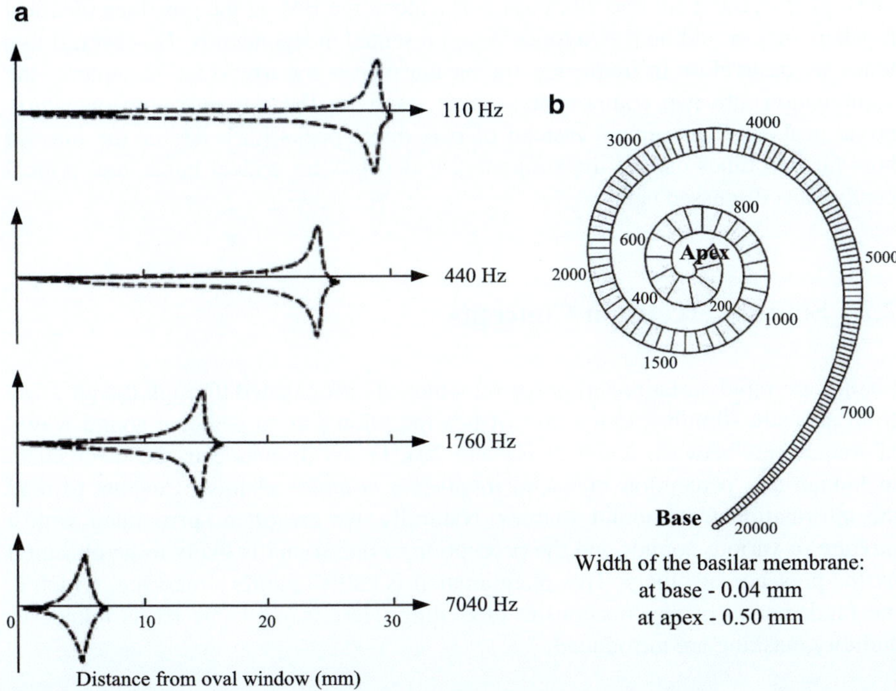

Fig. 2.4 Resonant properties of the basilar membrane (**a**) Envelopes of vibration patterns on the basilar membrane in response to sound of different frequencies [66] (**b**) Distribution of resonant frequencies along the basilar membrane [64]

summarizes the distribution of frequencies that produce maximum displacement at different positions along the basilar membrane. Note that the spacing of resonant frequencies is not linear to the frequency, but in a logarithmic scale approximately. It is called Bark scale or critical band rate corresponding to the concept of critical bands.

In this sense, the cochlea performs a transformation that maps sound frequencies onto certain locations along the basilar membrane, i.e., a "frequency-to-place" conversion [51]. It is of great importance to the comprehension of auditory masking. Since one frequency maximally excite only one particular point on the basilar membrane, the auditory system acts as a frequency analyzer which can distinguish the frequencies from each other. If two tones are different enough in frequency, the response of the BM to their combination is simply the addition of two individual ones. That is, there are two vibration peaks along the BM, at the positions identical to where they would be if two tones were presented independently. However, if two tones are quite close in frequency, the basilar membrane would fail to separate the combination into two components, which results in the response with one fairly broad peak in displacement instead of two single peaks [58]. As for the interval how far two tones can be discriminated, it depends on critical bands and critical bandwidths discussed next.

2.2 Sound Perception Concepts

Sounds are rapid variations in pressure, which are propagated through the air away from acoustic stimulus. Our sense of hearing allows us to perceive sound waves of frequencies between about 20 Hz and 20 kHz. As discussed in the mechanism of human ear, perception of sound involves a complex chains of events to read the information from sound sources. Naturally, we are often surrounded with a mixture of various sounds and the perception of one sound is likely to be obscured by the presence of others. This phenomenon is called auditory masking, which is the fundamental of psychoacoustic modelling. Here, some basic terms related to auditory masking are introduced.

2.2.1 Sound Pressure Level and Loudness

Sound reaches human ear in the form of pressure waves varying in time, $s(t)$. Physically, the pressure p is defined as force per unit area, and the unit in MKS system is Pascal (Pa) where $1\,\text{Pa} = 1\,\text{N/m}^2$. Also, the intensity is defined as power per unit area and its unit is W/m^2. In psychoacoustics, values of sound pressure vary from 10^{-5} Pa (ATH, absolute threshold of hearing) to 10^2 Pa (threshold of pain). To cover such a broad range, (SPL) is defined in logarithm units (dB) as

$$L_{\text{SPL}}/\text{dB} = 10\,\log_{10}\left(\frac{p}{p_0}\right)^2 = 10\,\log_{10}\left(\frac{I}{I_0}\right), \tag{2.1}$$

where L_{SPL} is the SPL of a stimulus, p is the pressure of stimulus in Pa, $p_0 = 20\,\mu\text{Pa}$ is the reference pressure of a tone with frequency around 2 kHz, I is sound intensity of the stimulus, and $I_0 = 10^{-12}\,\text{W/m}^2$ is the reference's intensity correspondingly [11].

The hearing sensation that relates to SPL is loudness of sound, expressed in *phon*. Note that loudness is a psychological, not a physical, attribute of sound. By definition, the loudness level of a 1 kHz tone is equal to its SPL in dB SPL [61]. The perceived loudness of sound depends upon its frequency as well as its intensity, as described by a series of equal-loudness contour in Fig. 2.5. Each equal-loudness

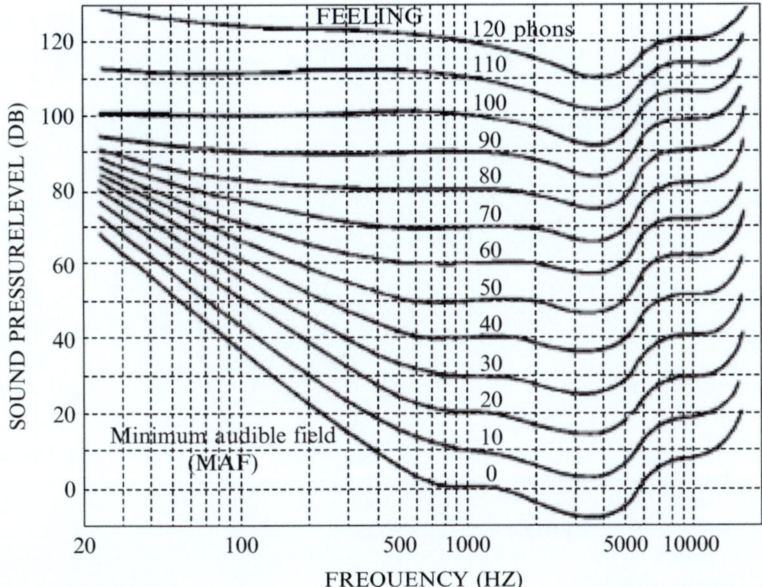

Fig. 2.5 Equal-loudness contours [69]

contour represents SPLs required at different frequencies in order that all tones on the contour are perceived equally loud [68]. The loudness of 20 phons contour at 100 Hz with 50 dB SPL is perceived similar to 1 kHz with 20 dB SPL. In Fig. 2.5, the deviation from the maximum sensitivity region of equal-loudness contours at high phons (i.e., 120 phons) is lower than those of low phons (i.e., 10 phons). This indicates that the sensitivity to frequency changing of HAS at low phons is relatively higher than high phons. Hence, complex sounds with identical frequency and phase components might sound different due to variations in loudness [58].

2.2.2 Hearing Range and Threshold in Quiet

Human hearing spreads widely from 20 Hz to 20 kHz in frequency, as well as ranging from about 0 dB up to 120 dB in SPL. The most sensitive part is between 100 Hz and 8 kHz for human speech. Figure 2.6 shows hearing range of human, where different hearing thresholds are sketched in SPL curves as function of frequency.

The hearing threshold at the bottom is the threshold in quiet, or (ATH), which approximately corresponds to the baseline in Fig. 2.5. It decreases gradually from 20 Hz to 3 kHz and then increases sharply above 16 kHz. The threshold in quiet indicates, as a function of frequency, the minimum SPL of a pure tone to be audible in a noiseless environment. Thus under no circumstances the human ear can perceive

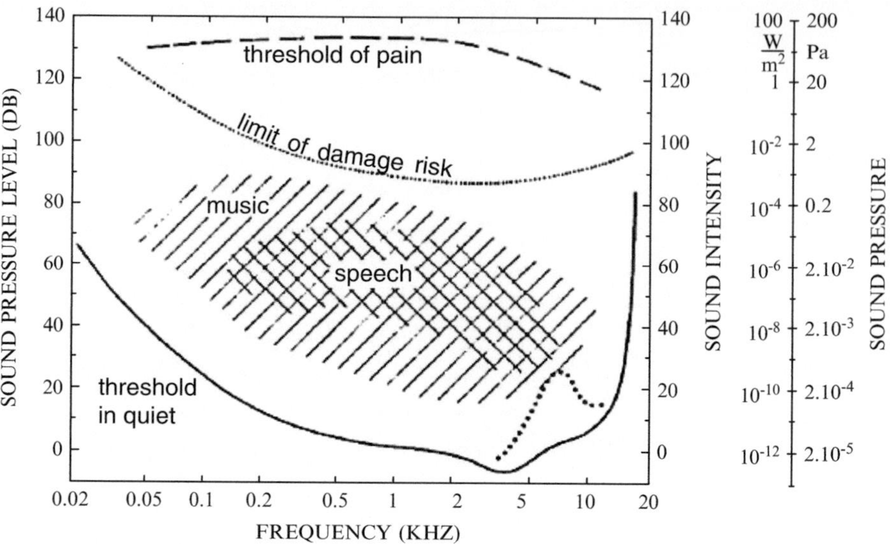

Fig. 2.6 Hearing range [11]

sounds at SPLs below that threshold. In other words, frequency components that fall below the threshold in quiet are insignificant to our perception of sound and unnecessary to be processed [51]. This property is crucial to the development of psychoacoustic model, where the threshold in quiet is approximated by the following frequency-dependent function:

Threshold in Quiet $(f)\,/\mathrm{dB} =$

$$3.64\left(\frac{f}{1000}\right)^{-0.8} - 6.5\exp\left\{-0.6\left(\frac{f}{1000} - 3.3\right)^2\right\} + 10^{-3}\left(\frac{f}{1000}\right)^4 \quad (2.2)$$

as plotted on both linear and logarithmic scales in Fig. 2.7. Regarding Eq. (2.2), one point to note is that it only applies to the frequency range $20\,\mathrm{Hz} \leq f \leq 20\,\mathrm{kHz}$.

2.2.3 Critical Bandwidth

As discussed in Sect. 2.1.3, the cochlea performs a "frequency-to-place" conversion and each position on the basilar membrane responds to a limited range of frequencies. Accordingly, the peripheral auditory system acts as a spectrum analyzer, modelling as a bank of band-pass filters with overlapping passbands [61]. Empirically, the main hearing range between $20\,\mathrm{Hz}$ and $16\,\mathrm{kHz}$ is divided into 24 nonoverlapping critical bands, and the critical bandwidths (CB) are listed

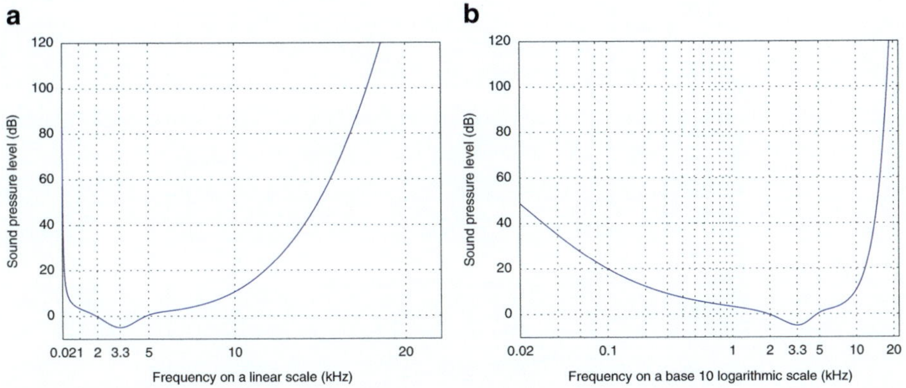

Fig. 2.7 Approximation for the threshold in quiet (**a**) Frequency on a linear scale (**b**) Frequency on a logarithmic scale

in Appendix D. We call it critical band rate scale and its unit is *Bark*. One Bark represents one critical band and corresponds to a distance along the basilar membrane of about 1.3 mm.[6] Considering nonlinear spacing of resonant frequencies on the basilar membrane, it is expected that critical bandwidths are nonuniform, varying as a function of frequency. The following equation describes the dependence of Bark scale on frequency [11]:

$$z/\text{Bark} = 13 \arctan\left(\frac{0.76 f_l}{1000}\right) + 3.5 \arctan\left(\frac{f_l}{7500}\right)^2, \qquad (2.3)$$

where f_l is the lower frequency limit of critical bandwidth. For example, the threshold in quiet in Fig. 2.7 is plotted on Bark scale as shown in Fig. 2.8.

Note that each critical bandwidth only depends on the center frequency of the passband. It is demonstrated in Fig. 2.9, where the critical bandwidth at 2 kHz is measured. As shown in Fig. 2.9a, hearing threshold is flat about 33 dB until two tones are about 300 Hz away from each other, and then it drops off rapidly. A similar result is obtained from Fig. 2.9b, hearing threshold is rather flat about 46 dB until two noises are away from 300 Hz [51]. Consequently, the critical bandwidth is 300 Hz for a center frequency of 2 kHz. It is worth mentioning that the threshold in Fig. 2.9b is at 46 dB versus only 33 dB in a, which means narrowband noises reduce

---

[6]The whole length of 32 mm basilar membrane divided by 24 critical bands is 1.3 mm for each band.

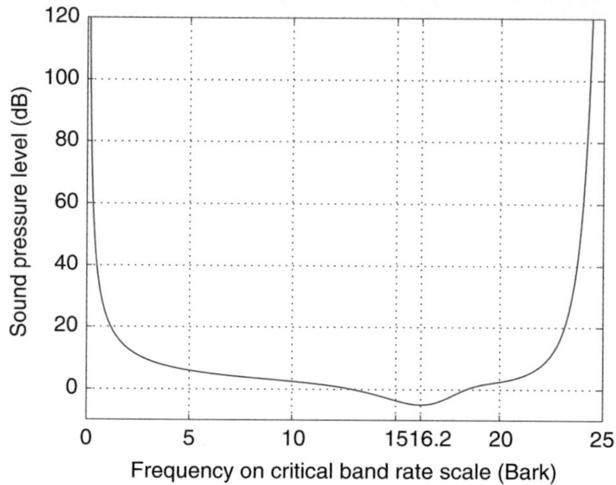

Fig. 2.8 Threshold in quiet on Bark scale

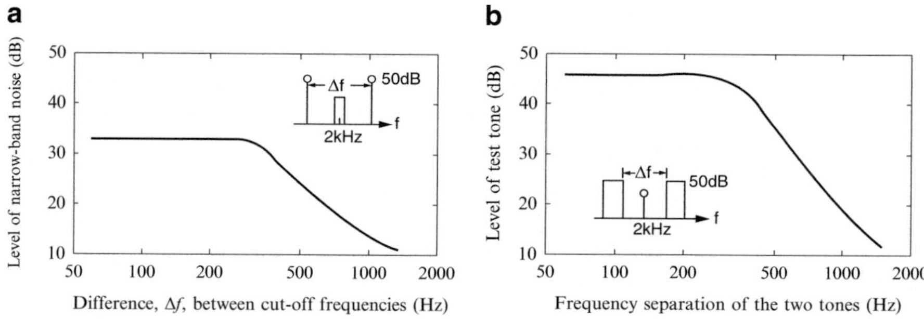

Fig. 2.9 Determination of the critical bandwidth [11] (a) The threshold for a narrowband noise 2 kHz centered between two tones of 50 dB as a function of the frequency separation between two tones (b) The threshold for a tone of 2 kHz centered between two narrowband noises of 50 dB as a function of the frequency separation between the cutoff frequencies of two noises

more audibility than tones. This fact is referred to "asymmetry of masking" and more details will be discussed in the next section.

On the basis of experimental data, an analytic expression is derived to better describe critical bandwidth $\triangle f$ as a function of center frequency f_c [11]:

$$\triangle f / \mathrm{Hz} = 25 + 75 \left[1 + 1.4 \left(\frac{f_c}{1000} \right)^2 \right]^{0.69} . \tag{2.4}$$

The concept of critical bandwidth contributes to the understanding of auditory masking, because CB around a masker denotes the frequency range over which the main masking effect operates. As demonstrated in Sect. 2.3.1, the masking curves distribute equally across the spectrum in Bark scale.

2.3 Auditory Masking

Due to the effect of auditory masking, the perception of one sound is related to not only its own frequency and intensity, but also its neighbor components. Auditory masking refers to the phenomenon that one faint but audible sound (the maskee) becomes inaudible in the presence of another louder audible sound (the masker). It has a great influence on hearing sensation and involves two types of masking, i.e., simultaneous masking and nonsimultaneous masking (including pre-masking and post-masking) as displayed in Fig. 2.10. Due to auditory masking, any signals below these curves cannot be heard. Therefore, by virtue of auditory masking, we can modify audio signals in a certain way without perceiving deterioration, as long as the modifications could be properly "masked." This notion is the essence of audio watermarking [70, 71].

2.3.1 Simultaneous Masking

Simultaneous masking (or frequency masking) refers to masking between two sounds with close frequencies, where the low-level maskee is made inaudible by simultaneously occurring louder masker. Both masker and maskee can be sinusoidal

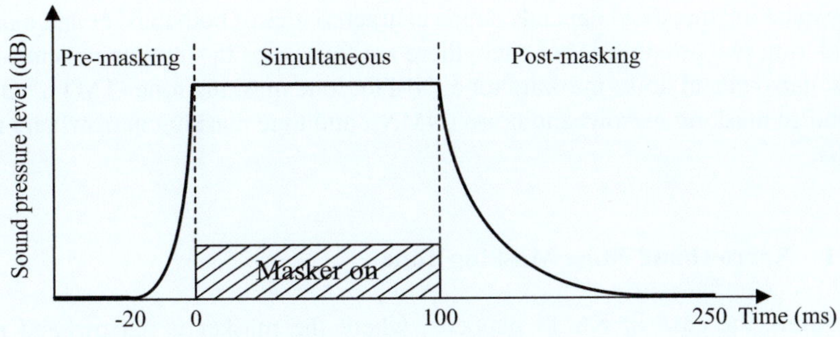

Fig. 2.10 Two types of masking: simultaneous and nonsimultaneous masking

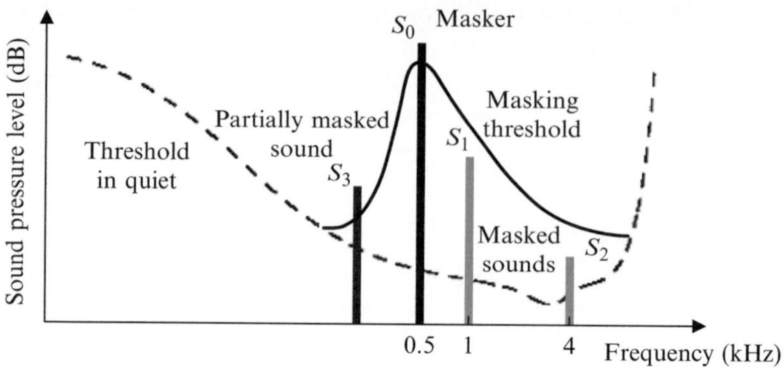

Fig. 2.11 Simultaneous masking

tone or narrowband noise.[7] Figure 2.11 gives an example of simultaneous masking, where sound S_0 is the masker. Because of the presence of S_0, the threshold in quiet is elevated to produce a new hearing threshold named as masking threshold. The masking threshold is a kind of limit for just noticeable distortion (JND) [72], which means that any sounds or frequency components below this threshold are masked by the presence of the masker. For instance, the weaker signal S_1 and S_2 are completely inaudible, as their SPLs are below the masking threshold. For the signal S_3, it is partially masked and only the portion above the threshold is perceivable. Moreover, the effective masking ranges for the maskers at different frequencies are determined solely by critical bandwidths, as implied in Fig. 2.9. If the maskee lies in critical band of the masker, the maskee is more likely to be unperceived. The mechanism by which masking occurs is still uncertain [61]. In general, it is because the louder masker creates an excitation of sufficient strength on the basilar membrane. Then such an excitation prevents the detection of another excitation within the same critical band from a weaker sound [51].

The masking threshold depends on the characteristics of both masker and maskee. Considering two possibilities of each, there are four cases in simultaneous masking, that is, narrowband noise masking tone (NMT), tone masking tone (TMT), narrowband noise masking narrowband noise (NMN), and tone masking narrowband noise (TMN).

2.3.1.1 Narrowband Noise Masking Tone

Most often, the case of NMTs happens, where the masker is narrowband noise and the maskees are tones located in the same critical band. Figure 2.12 shows the masking thresholds for narrowband noise masker masking tones, where the noise is

---

[7]Here, narrowband means the bandwidth equal to or smaller than a critical band.

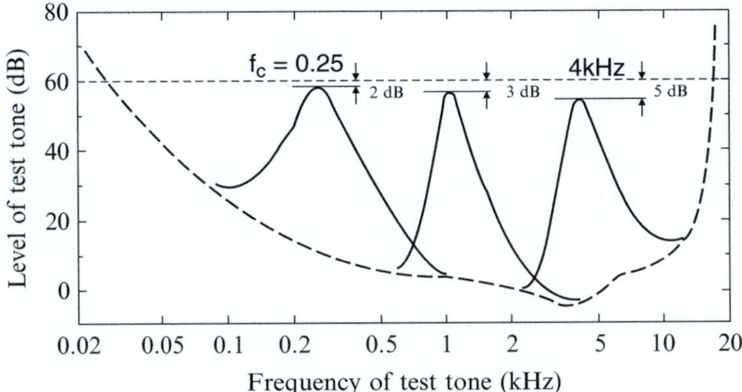

Fig. 2.12 Masking thresholds for a 60 dB narrowband noise masker centered at different frequencies [51]

at a SPL of 60 dB and centered at 0.25, 1, and 4 kHz separately. In the graph, solid lines represent masking thresholds, and the dashed line at the bottom is the threshold in quiet.[8] The masking thresholds have a number of important features. For example, the form of curve varies with different maskers, but always reaches a maximum near the masker's center frequency. It means that the amount of masking is greatest when the maskee is located at the same frequency with the masker. The masking ability of a masker is indicated by the minimum signal to mask ratio (SMR), i.e., the minimum difference of SPL between the masker and its masking threshold. Therefore, higher SMR implies less masking. Another point is that low-frequency masker produces a broader masking threshold and provides more masking than high frequencies. Here, the 0.25, 1, and 4 kHz thresholds have a SMR of 2, 3, and 5 dB, respectively.

Figure 2.12 is sketched in normal frequency units, where the masking thresholds of different frequencies are dissimilar in shape. If graphed in Bark scale, all the masking thresholds look similar in shape as shown in Fig. 2.13.[9] In this case, it is easier to model the masking threshold by the use of the so-called spreading function in Sect. 2.4.1.1. As a result, Bark scale is widely used in the area of auditory masking.

Moreover, the masking thresholds from a 1 kHz narrowband noise masker at different SPLs, L_{CB}, are outlined in Fig. 2.14. Although SPL of the masker is different, the minimum SMR remains constant at around 3 dB, corresponding to the value in Fig. 2.12. It means that the minimum SMR in NMT solely depends on the center frequency of masker. Also notice that the masking threshold becomes more asymmetric around the center frequency as the SPL increases. At frequencies lower than 1 kHz, all the curves have a steep rise. But at frequencies higher than 1 kHz,

---

[8]Hereafter, this rule does apply to all the graphs in Sect. 2.3.

[9]For illustration, all the curves are shifted upward to the masker's SPL (60 dB).

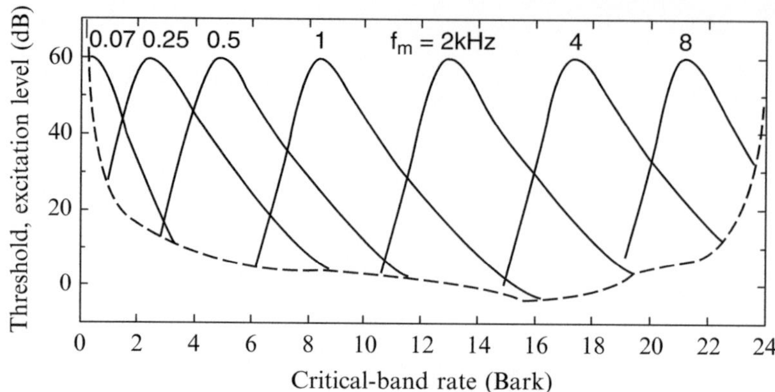

Fig. 2.13 Masking thresholds for a 60 dB narrowband noise masker centered at different frequencies in Bark scale [51]

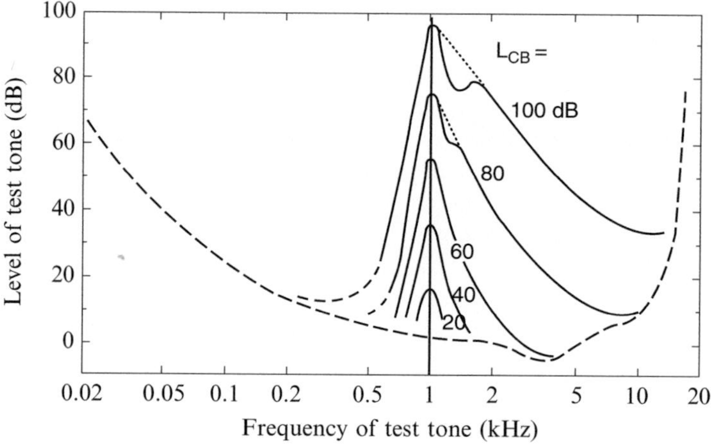

Fig. 2.14 Masking thresholds from a 1 kHz narrowband noise masker at different SPLs [51]

the slopes of maskers at higher SPLs decrease more gradually. Recall Fig. 2.4a; it is reasonable to expect that the masker is good at masking the tones whose frequencies are lower than its own frequency, rather than higher frequency tones [58]. To show the similarity in shape over all the masking thresholds, Fig. 2.15 plots the curves in Bark scale again.

2.3.1.2 Tone Masking Tone

The early work on auditory masking started from experiments on tones masking tones within the same critical band. Since both the masker and maskee are pure tones, their interference is likely to result in the occurrence of beats. Therefore,

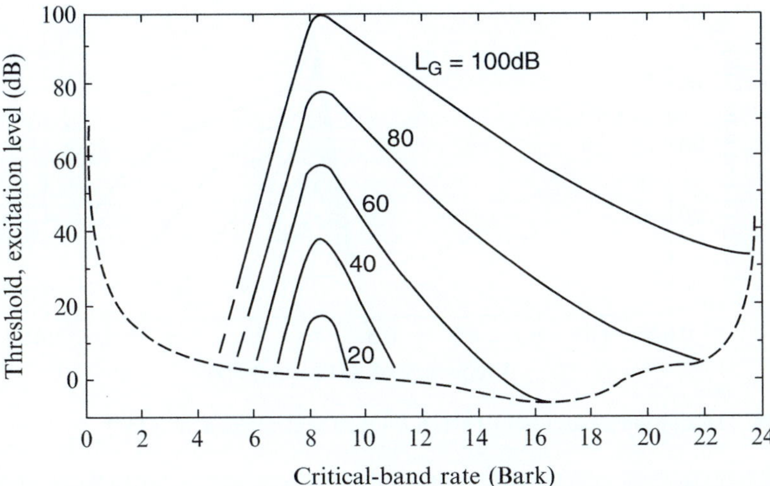

Fig. 2.15 Masking thresholds from a 1 kHz narrowband noise masker at different SPLs in Bark scale [51]

besides the masker and maskee, additional beating tones become audible and accordingly disturb the subjects. Figure 2.16 shows the masking thresholds from a 1 kHz tonal masker at different SPLs. During the course of approaching 1 kHz, the maskee was set 90° out of phase with the masker to prevent beating. Similar to Fig. 2.14, the masking thresholds spread also broader towards high frequencies than lower frequencies. However, an obvious difference lies in the minimum SMR, roughly 15 dB in Fig. 2.16 versus about 3 dB in Fig. 2.14. It indicates that the narrowband noise is a better masker than pure tone, referred as "asymmetry of masking" [73]. This fact actually has been demonstrated in Fig. 2.9 already. The masking threshold by narrowband noise masker in Fig. 2.9b is valued at 46 dB, higher than 33 dB by tonal masker in Fig. 2.9a. So in psychoacoustic modelling, we should identify the frequency components to be noise-like or tone -like and then calculate their masking thresholds separately.

2.3.1.3 Narrowband Noise or Tone Masking Narrowband Noise

In contrast to NMT and TMT, it is more difficult to characterize narrowband noise or tone masking narrowband noise. So far, relatively few studies in NMN and TMN are carried out. Under the case of NMN, the masking thresholds heavily rely on phase relationship between the masker and maskee. In other words, different relative phases between the masker and maskee would lead to different values of minimum SMRs. It is reported that measurements for wideband noise have minimum SMRs of about 26 dB [51,73]. As for TMN, the minimum SMR tends to fluctuate between 20 and 30 dB [51].

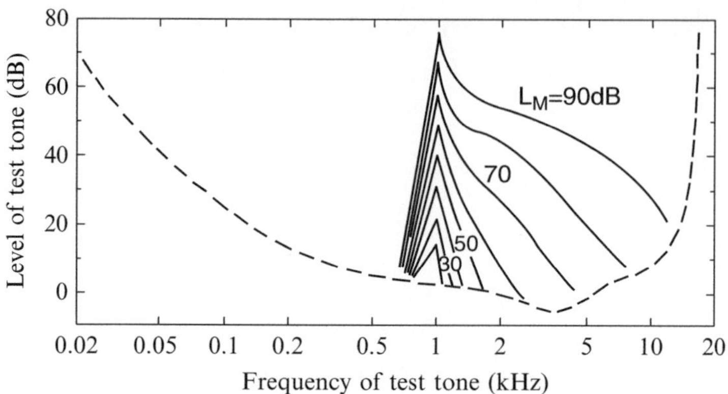

Fig. 2.16 Masking thresholds from a 1 kHz tonal masker at different SPLs [51]

2.3.2 Nonsimultaneous Masking

In addition to simultaneous masking, auditory masking can also take place when
the maskee is present immediately preceding or following the masker. This is called
nonsimultaneous masking or temporal masking. As exemplified in Fig. 2.10, one
200 ms masker masks a tone burst with very short duration relative to the masker.

There are two kinds of nonsimultaneous masking: (1) pre-masking or backward
masking, occurring just before the onset of masker, and (2) post-masking or forward
masking, occurring after the removal of masker. In general, the physiological
basis of nonsimultaneous masking is that the auditory system requires a certain
integration time to build the perception of sound, where louder sounds require longer
integration intervals than softer ones [51].

2.3.2.1 Pre-masking

Pre-masking is somewhat unexpected since it happens before the presence of
masker. As seen from Fig. 2.10, the duration of pre-masking is quite short (about
20 ms), whereas it is most effective only in 1–2 ms before the onset of masker [73].
It is suggested that the duration of masker might affect the time that pre-masking
lasts. Up to now, however, no experimental results could specify such a relation.

Pre-masking has less masking capacity than post-masking and simultaneous
masking; nevertheless, it plays a significant role in the compensation of pre-noise or
pre-echo distortion [51].

2.3.2.2 Post-masking

Post-masking is better understood compared to pre-masking. It reflects a moderate decrease of the masking level after the masker is halted. As displayed in Fig. 2.10, post-masking level decays gradually to zero after a longer period of time (about 150 ms). Therefore, post-masking exhibits a higher masking capacity which is beneficial to most applications. Experimental studies have revealed that post-masking depends on the intensity and duration of the masker as well as relative frequency of the masker and maskee [51].

2.4 Psychoacoustic Model

The knowledge of auditory masking provides the foundation for developing psychoacoustic models. In psychoacoustic modelling, we use empirically determined masking models to analyze which frequency components contribute more to the masking threshold and how much "noise" can be mixed in without being perceived. This notion is applicable to audio watermarking, of which the imperceptibility is one prerequisite. Typically, in some audio watermarking techniques such as spread spectrum watermarking [74, 75] and wavelet domain watermarking [7, 76], the watermark signal is added to the host signal as a faint additive noise. To keep the watermarks inaudible, we often utilize the minimum masking threshold (MMT) calculated from psychoacoustic model to shape the amplitude of watermark signal.

2.4.1 Modelling the Effect of Simultaneous Masking

Modelling the effect of simultaneous masking is one major task of psychoacoustic model. In general, there are a series of steps involved. Firstly, the input audio signal is analyzed to classify its noise-like and tone-like frequency components, due to the phenomenon of "asymmetry of masking." Secondly, the so-called spreading functions are derived to mimic the excitation patterns of noise-like and tone-like maskers, respectively. Thirdly, after shifted down by a certain amount for each masker, all the individual masking thresholds as well as ATH are added up in some manner to obtain a global masking threshold, an estimation on the concurrent masking effect. Finally, we take the lowest level of global masking threshold in each frequency band to obtain the (MMT), which represents the most sensitive limit.

2.4.1.1 Models for the Spreading of Masking

Models for the spreading of masking are developed to delineate excitation patterns of the maskers. As noticed from two examples of excitation patterns in Figs. 2.13

and 2.15, the shape of curves are quite similar and also easy to describe in Bark scale, because Bark scale is linearly related to basilar membrane distances. Accordingly, we define spreading function SF(dz) as a function of the difference between the maskee and masker frequencies in Bark scale, $dz/\text{Bark} = z(f_{\text{maskee}}) - z(f_{\text{masker}})$. Apparently, $dz \geq 0$ when the masker is located at a lower frequency than the maskee, and $dz < 0$ when the masker is located at a higher frequency than the maskee.

There are a number of spreading functions introduced to imitate the characteristics of maskers. For instance, two-slope spread function is the simplest one that uses a triangular function:

$$10 \log_{10} \text{SF} (dz) /\text{dB} = \begin{cases} [-27 + 0.37 \max \{L_M - 40, 0\}] \, dz, & dz \geq 0 \\ 27 dz, & dz < 0, \end{cases} \tag{2.5}$$

where L_M is SPL of the masker.

Another popular spreading function is proposed by Schroeder and expressed as the following analytical function:

$$10 \log_{10} \text{SF} (dz) /\text{dB} = 15.81 + 7.5 (dz + 0.474) - 17.5 \sqrt{1 + (dz + 0.474)^2}. \tag{2.6}$$

After slight modification on Schroeder's spreading function, spreading function as Eq. (2.7) is adopted in ISO/IEC MPEG[10] Psychoacoustic Model 2.

$$10 \log_{10} \text{SF} (dz) /\text{dB}$$
$$= 15.8111389 + 7.5 (1.05 dz + 0.474) - 17.5 \sqrt{1 + (1.05 dz + 0.474)^2}$$
$$+ 8 \min \left(0, \left[(1.05 dz - 0.5)^2 - 2 (1.05 dz - 0.5) \right] \right). \tag{2.7}$$

It should be noted that the two spreading functions Eqs. (2.6) and (2.7) are independent of the masker's SPL, which is advantageous to reduction in computation when generating overall masking threshold.

The spreading function utilized in ISO/IEC MPEG Psychoacoustic Model 1 is different from Psychoacoustic Model 2:

$$10 \log_{10} \text{SF} (dz) /\text{dB} = \begin{cases} 17 dz - 0.4 L_M + 11, & -3 \leq dz < -1 \\ (0.4 L_M + 6) \, dz, & -1 \leq dz < 0 \\ -17 dz, & 0 \leq dz < 1 \\ -17 dz + 0.15 L_M (dz - 1), & 1 \leq dz < 8 \end{cases}. \tag{2.8}$$

---

[10]ISO: International Organization for Standardization; IEC: International Electrotechnical Committee; MPEG: Moving Picture Experts Group.

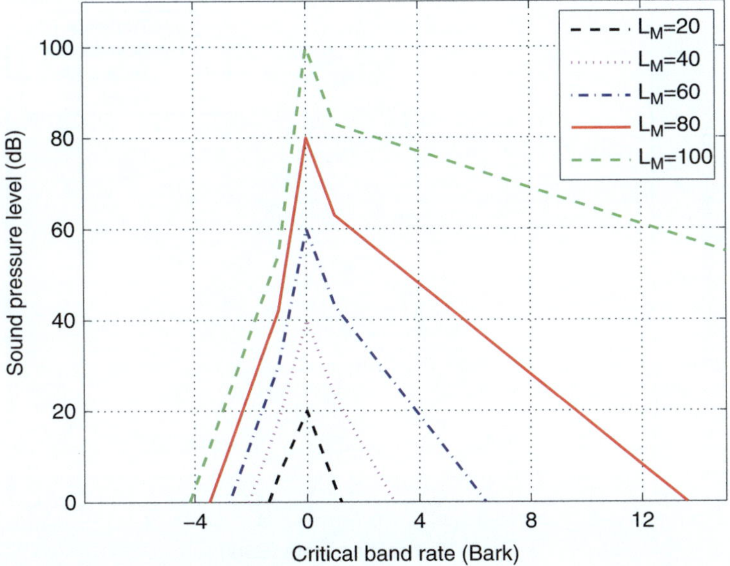

Fig. 2.17 Spreading function in ISO/IEC Psychoacoustic Model 1

Figure 2.17 shows spreading functions in Model 1 for different levels of the masker. It is seen that the higher SPL the masker has, the more asymmetric the curve looks. Specifically, higher frequencies exhibit more masking than lower frequencies when the level of masker is high. This two-piece linear spreading function is a good approximation to the masking thresholds of TMT in Fig. 2.16.

In addition, four models described above for spreading functions, i.e., two-slope SF, Schroeder SF, Psychoacoustic Model 1 SF, and Model 2 SF, are compared at a level of 80 dB in Fig. 2.18. Among these four models, two-slope spreading function is the most conservative one, and Model 1 spreading function allows for more upward spreading of masking than others [51].

2.4.1.2 Implementation of Psychoacoustic Model 1

In different application scenarios, psychoacoustic model can be implemented in different ways to satisfy the criteria required. ISO/IEC MPEG-1 Standard [77] utilizes two informative psychoacoustic models, Psychoacoustic Model 1 and 2, to determine the MMT for inaudibility. Typically, Model 1 is applied to MPEG Layers I and II and Model 2 to MPEG Layer III. Both models are commonly in use and well performed. Psychoacoustic Model 1 proposed a low-complication method to analyze spectral data and output SMR, whereas Psychoacoustic Model 2 performs a more detailed analysis at the expense of greater computational complexity

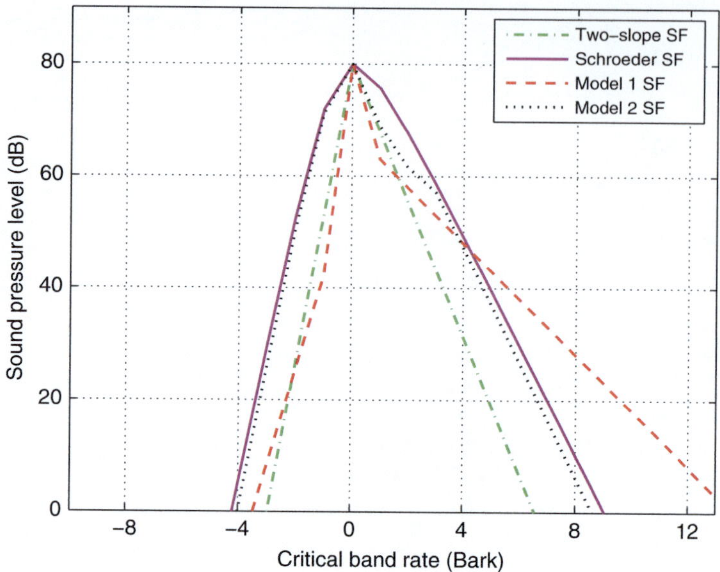

Fig. 2.18 Comparison of four spreading functions relative to an 80 dB masker

[31, 78, 79]. Hence, Psychoacoustic Model 1 for Layer I is later employed in our audio watermarking scheme in consideration of its higher efficiency.

In our case, the input to Psychoacoustic Model 1 is one frame of audio signal and the corresponding output is its MMT. The whole procedure of implementation consists of six steps [72, 73, 77, 80]:

1. FFT analysis and SPL normalization
2. Identification of tonal and nontonal maskers
3. Decimation of invalid tonal and nontonal maskers
4. Calculation of individual masking thresholds
5. Calculation of global masking threshold
6. Determination of the MMT

The details of each step are expounded as follows:

- **STEP 1:** FFT analysis and SPL normalization

For an accurate analysis of frequency components, fast Fourier transform (FFT) is performed to obtain a high-resolution spectral estimate of incoming frame $x(n)$. In Psychoacoustic Model 1, the input frame has a size of $N = 512$ points. To minimize the leakage effect, $x(n)$ is multiplied with a modified Hanning window $w(n)$ defined by

$$w(n) = \sqrt{\frac{8}{3}}\text{hann}(N) = \sqrt{\frac{8}{3}} \cdot \frac{1}{2}\left[1 - \cos\left(\frac{2\pi n}{N}\right)\right] \quad 0 \le n \le N - 1, \quad (2.9)$$

where $\mathrm{hann}\,(N) = \frac{1}{2}\left[1 - \cos\left(\frac{2\pi n}{N}\right)\right]$ is the N-point Hanning window. Factor $\sqrt{\frac{8}{3}}$ is a gain to compensate the average power of $w\,(n)$, so that $\langle w\,(n)^2\rangle \equiv \frac{1}{N}\sum_{n=0}^{N-1}\left[w\,(n)^2\right] = 1$. Then, power spectral density (PSD) of $x\,(n)$ is computed as

$$\mathrm{PSD}\,(k)\,/\mathrm{dB} = 10\log_{10}\left|\frac{1}{N}\left[\sum_{n=0}^{N-1} x\,(n)\,w\,(n)\exp\left(-j\frac{2\pi nk}{N}\right)\right]\right|^2 \quad 0 \le k < \frac{N}{2}.$$

(2.10)

After that, PSD estimate $\mathrm{PSD}\,(k)$ is normalized to a SPL level of 96 dB, i.e., the maximal is limited to 96 dB.

$$\begin{aligned} P\,(k)\,/\mathrm{dB} &= 96 - \max\{\mathrm{PSD}\,(k)\} + \mathrm{PSD}\,(k) \\ &= \triangle_P + \mathrm{PSD}\,(k), \end{aligned}$$

(2.11)

where $\triangle_P = 96 - \max\{\mathrm{PSD}\,(k)\}$. It is because we have no prior knowledge regarding actual playback levels, the absolute pressure level of a sound can only be specified by comparing to a reference. To this end, a sinusoid with amplitude equal to half of PCM quantizer spacing $\left(A_0 = \frac{\triangle}{2}\right)$ is defined as having a SPL of 0 dB, i.e., $20\log_{10}\left(A_0/A_0\right) = 0\,\mathrm{dB}$. Consequently, for 16-bit PCM data, a sinusoid with amplitude equal to the overload level of quantizer $\left(A_{\max} = \frac{\left(2^{16}-1\right)\triangle}{2}\right)$ would have a SPL of about 96 dB, i.e., $20\log_{10}\left(A_{\max}/A_0\right) = 20\log_{10}\left(2^{16} - 1\right) \approx 96\,\mathrm{dB}$ [51].

An example of the initial and normalized PSD estimates as well as the threshold in quiet are shown in Fig. 2.19, where the frequencies of two graphs are plotted on linear and Bark scales, respectively. Note that in psychoacoustic models, an offset depending on the overall bit rate is employed for the threshold in quiet. It is equal to $-12\,\mathrm{dB}$ for bit rates no less than 96 kbits/s and 0 dB for bit rates less than 96 kbits/s per channel [77]. Sound tracks used in our experiments are of CD quality, whose bit rates are normally greater than 96 kbits/s. Therefore, by comparing Fig. 2.19 to Figs. 2.7 and 2.8, the threshold in quiet in Fig. 2.19 is shifted downward by 12 dB.

- **STEP 2:** Identification of tonal and nontonal maskers

On account of "asymmetry of masking," it is required to discern frequency components as tonal (i.e., sinusoidal) and nontonal (i.e., noise-like) maskers. Tonal maskers are selected from local maxima of normalized PSD estimate, $P\,(k)$. A local maxima refers to the maximum PSD within its two neighbors:

$$P\,(k) \ge P\,(k + 1) \quad \text{and} \quad P\,(k) \ge P\,(k - 1)$$

(2.12)

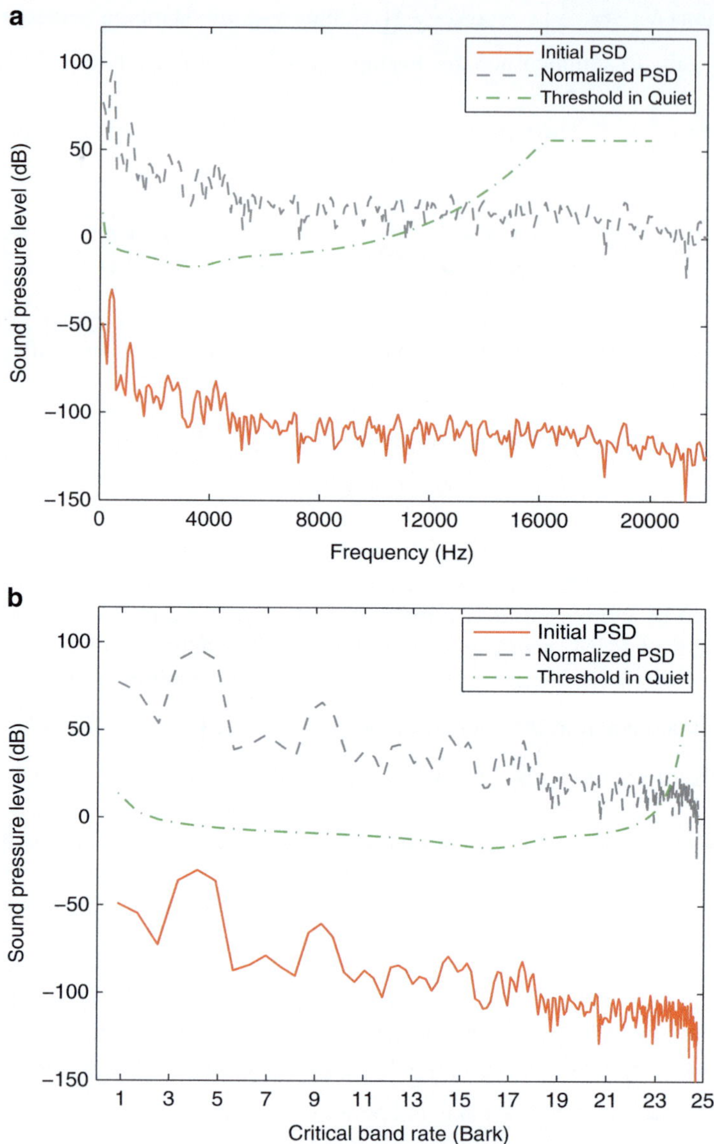

Fig. 2.19 Initial and normalized PSD estimates (**a**) Frequency on linear scale (**b**) Frequency on
Bark scale

If the value of a local maxima is at least 7 dB greater than that of its neighboring components within a certain Bark range D_k, such a maxima will be marked as a tonal masker. All the tonal components comprise the "tonal" set, S_{TM}:

$$S_{TM} = \{P\,(k) \mid [P\,(k) - P\,(k \pm D_k)] \geq 7\,\text{dB}\}, \qquad (2.13)$$

where D_k varies with different frequency indices.[11]

$$D_k \in \begin{cases} \{\pm 2\}, & 2 < k < 63 & \leftrightarrow \frac{2F_s}{N} \sim \frac{63F_s}{N}\,\text{kHz} \\ \{\pm 2, \pm 3\}, & 63 \leq k < 127 & \leftrightarrow \frac{63F_s}{N} \sim \frac{127F_s}{N}\,\text{kHz} \\ \{\pm 2, \pm 3, \ldots, \pm 6\}, & 127 \leq k \leq 250 & \leftrightarrow \frac{127F_s}{N} \sim \frac{250F_s}{N}\,\text{kHz} \end{cases}.$$

One point to note is that [77] did not specify the value of D_k for $251 \leq k \leq 256$, because the maskers within this range are already dominated by the threshold in quiet (as seen in Fig. 2.19) and have no contribution to masking threshold. Actually, it is the first criterion for decimation in Step 3.

As the effect of masking is additive in the logarithmic domain, the SPL of each tonal component is calculated by

$$P_{TM}\,(k)\,/\text{dB} = 10\log_{10}\left[10^{\frac{P(k-1)}{10}} + 10^{\frac{P(k)}{10}} + 10^{\frac{P(k+1)}{10}}\right]. \qquad (2.14)$$

In addition, the remaining components within each critical band[12] are treated to be nontonal. So we sum up their intensities as the SPL of a single nontonal masker for each critical band, P_{NM}:

$$P_{NM}\left(\overline{k}\right)/\text{dB} = 10\log_{10}\sum_j\left[10^{\frac{P(j)}{10}}\right] \quad \forall P\,(j) \notin S_{TM}, \qquad (2.15)$$

where that $\overline{k}$ is the frequency index nearest to the geometric mean[13] of each critical band. Correspondingly, all the nontonal components are put into the "nontonal" set, S_{NM}.

---

[11] The frequency edges are calculated based on the sampling frequency F_s.

[12] Critical band boundaries vary with the Layer and sampling frequency. ISO/IEC IS 11172-3 [77] has tabulated such parameters in Table D.2a–f. In our case, Table D.2b for Layer I at a sampling frequency of 44.1 kHz is adopted.

[13] The geometric mean of a data set $[a_1, a_2, \ldots, a_M]$ is defined as $\left(\prod_{m=1}^{M} a_m\right)^{1/M}$. It is sometimes called the log-average, i.e., $\left(\prod_{m=1}^{M} a_m\right)^{1/M} = 10^{\wedge}\left[\frac{1}{M}\sum_{m=1}^{M}\log_{10}(a_m)\right]$.

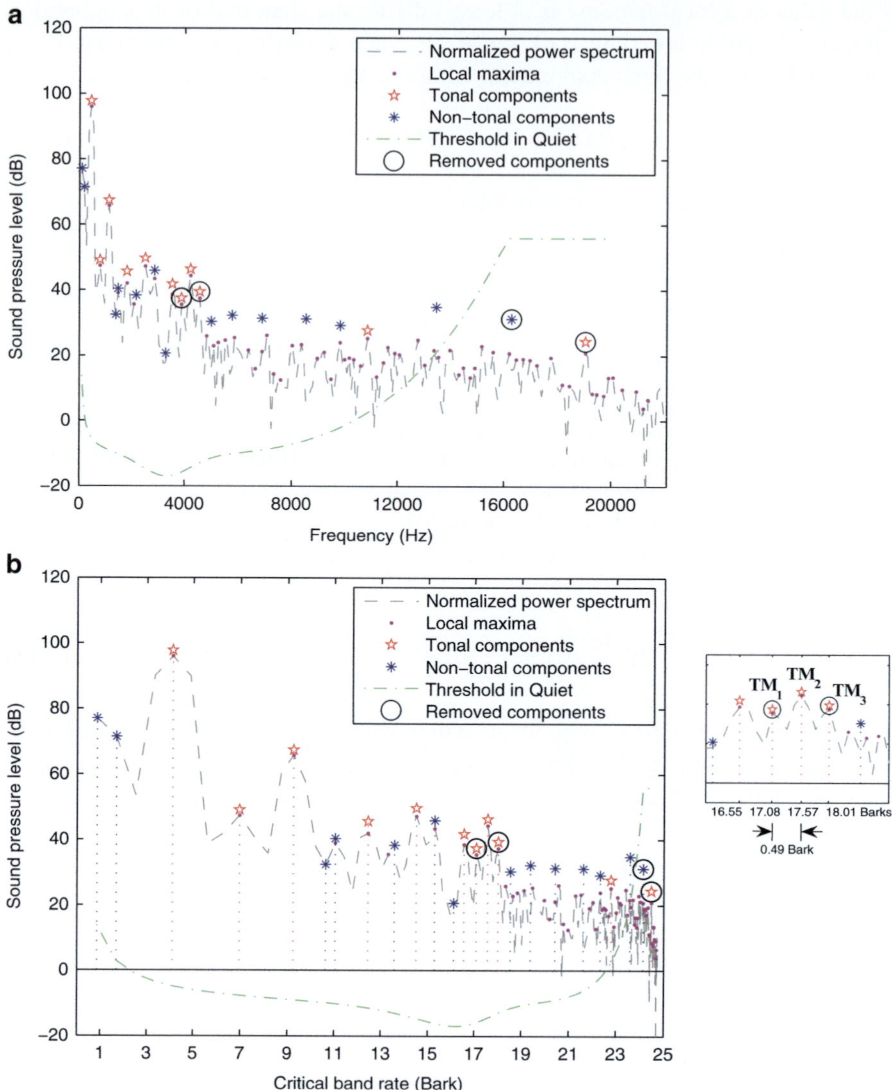

Fig. 2.20 Tonal and nontonal maskers (**a**) Frequency on a linear scale (**b**) Frequency on Bark scale

Tonal and nontonal maskers are denoted by pentagram and asterisk symbols in Fig. 2.20, respectively. Particularly, the associated critical band for each masker is indicated in the graph on Bark scale.

- **STEP 3:** Decimation of invalid tonal and nontonal maskers

On considering their possible contributions to masking threshold, the sets of tonal and nontonal maskers are examined according to two criteria as follows:

One rule is that any tonal and nontonal maskers below the threshold in quiet are removed. That is, only the maskers that satisfy Eq. (2.16) are retained, where ATH (k) is the SPL of threshold in quiet at frequency index k:

$$P_{\text{TM, NM}}(k) \geq \text{ATH}(k).\tag{2.16}$$

For example, one of each tonal and nontonal maskers between 24 and 25 Barks is discarded, as shown in Fig. 2.20b.

The other rule is to simplify any group of maskers occurring within a distance of 0.5 Bark: only the masker with the highest SPL is preserved and the rest are eliminated.

$$P_{\text{TM, NM}}(k) = \underset{k_0 \in [-0.5,0.5]}{\arg \max} \; P_{\text{TM, NM}}(k + k_0).\tag{2.17}$$

For example, two pairs of tonal maskers between 17 and 19 Barks, $\{\text{TM}_1, \text{TM}_2\}$ and $\{\text{TM}_2, \text{TM}_3\}$, are inspected. As shown in an enlarged drawing on the right of Fig. 2.20b, the distance between $\{\text{TM}_1, \text{TM}_2\}$ is 0.49 Bark, and TM_1 has a lower SPL than TM_2. Therefore, TM_2 is preserved, whereas TM_1 is removed. Similarly, we dispose of TM_3 but retain TM_2 for $\{\text{TM}_2, \text{TM}_3\}$.

$$\text{TM}_2 \longleftarrow \{\text{TM}_1, \text{TM}_2\} \left| \begin{array}{l} \text{Distance} : 17.57 - 17.08 = 0.49 \, \text{Bark} \\ \text{SPL} : \; P_{\text{TM}_1} < P_{\text{TM}_2} \end{array} \right.$$

$$\text{TM}_2 \longleftarrow \{\text{TM}_2, \text{TM}_3\} \left| \begin{array}{l} \text{Distance} : 18.01 - 17.57 = 0.44 \, \text{Bark} \\ \text{SPL} : \; P_{\text{TM}_2} > P_{\text{TM}_3} \end{array} \right.$$

In Fig. 2.20, the invalid tonal and nontonal maskers being decimated are denoted by a circle.

- **STEP 4:** Calculation of individual masking thresholds

After eliminating invalid maskers, individual masking threshold is computed for each tonal and nontonal masker. An individual masking threshold $L(j, i)$ refers to the masker at frequency index j contributing to masking effect on the maskee at frequency index i. It corresponds to $L[z(j), z(i)]$, where $z(j)$ and $z(i)$ are the masker and maskee's frequencies in Bark scale. In MPEG psychoacoustic models, only a subset of samples over the whole spectrum are considered to be maskees and involved in the calculation of global masking threshold. The number and frequencies of maskees also depend on the Layer and sampling frequency, as tabulated from Table D.1a–f in [77]. In our case, Table D.1b for Layer I at a sampling frequency of 44.1 kHz is adopted, where 106 maskees are taken into account.

The individual masking thresholds for tonal and nontonal maskers, $L_{\text{TM}}[z(j), z(i)]$ and $L_{\text{NM}}[z(j), z(i)]$, are calculated by

$$L_{\text{TM}}[z(j), z(i)]/\text{dB} = P_{\text{TM}}[z(j)] + \triangle_{\text{TM}}[z(j)] + \text{SF}[z(j), z(i)]\tag{2.18}$$

$$L_{\mathrm{NM}}\left[z\left(j\right),z\left(i\right)\right]/\mathrm{dB} = P_{\mathrm{NM}}\left[z\left(j\right)\right] + \triangle_{\mathrm{NM}}\left[z\left(j\right)\right] + \mathrm{SF}\left[z\left(j\right),z\left(i\right)\right], \quad (2.19)$$

where $P_{\mathrm{TM}}\left[z\left(j\right)\right]$ and $P_{\mathrm{NM}}\left[z\left(j\right)\right]$ are the SPLs of tonal and nontonal maskers at a Bark scale of $z\left(j\right)$, respectively. The term $\triangle_X$ is called masking index, an offset between the excitation pattern and actual masking threshold. As mentioned in Sect. 2.3.1, the excitation pattern needs to be shifted by an appropriate amount in order to obtain the masking curve relative to the masker. Because tonal and nontonal maskers have different masking capability, i.e., the noise is a better masker than pure tone, the masking indices of tonal and nontonal maskers are defined separately as follows [77]:

$$\triangle_{\mathrm{TM}}\left[z\left(j\right)\right] = -6.025 - 0.275z\left(j\right) \qquad (2.20)$$

$$\triangle_{\mathrm{NM}}\left[z\left(j\right)\right] = -2.025 - 0.175z\left(j\right). \qquad (2.21)$$

The term $\mathrm{SF}\left[z\left(j\right),z\left(i\right)\right]$ is the spreading function discussed already in Sect. 2.4.1.1. Psychoacoustic Model 1 employs spreading function in Eq. (2.8), rewritten in the following expression:

$$10\log_{10}\mathrm{SF}\left(dz\right)/\mathrm{dB} = \begin{cases} 17dz - 0.4P_X\left[z\left(j\right)\right] + 11, & -3 \leq dz < -1 \\ \left(0.4P_X\left[z\left(j\right)\right] + 6\right)dz, & -1 \leq dz < 0 \\ -17dz, & 0 \leq dz < 1 \\ -17dz + 0.15P_X\left[z\left(j\right)\right]\left(dz - 1\right), & 1 \leq dz < 8 \end{cases},$$

$$(2.22)$$

where dz is the distance from the maskee to masker, $dz = z\left(i\right) - z\left(j\right)$, as defined in Sect. 2.4.1.1. $P_X\left[z\left(j\right)\right]$ refers to $P_{\mathrm{TM}}\left[z\left(j\right)\right]$ in the case of tonal masker, otherwise $P_{\mathrm{NM}}\left[z\left(j\right)\right]$ for nontonal masker. Notice that for reasons of implementation complexity, the masking is no longer considered if $dz < -3$ Bark or $dz \geq 8$ Bark and thereby $L_{\mathrm{TM}}\left[z\left(j\right),z\left(i\right)\right]$ and $L_{\mathrm{NM}}\left[z\left(j\right),z\left(i\right)\right]$ are set to $-\infty$ dB outside the above ranges [77].

Figure 2.21 shows the individual masking thresholds for both tonal and nontonal maskers survived from the decimation.

- **STEP 5:** Calculation of global masking threshold

The global masking threshold is the combination of individual masking thresholds and the threshold in quiet. Since the mixture of masking is additive, the global masking threshold at frequency index i is calculated according to

$$L_G\left(i\right)/\mathrm{dB} = 10\log_{10}\left[10^{\frac{\mathrm{ATH}(i)}{10}} + \sum_{j=1}^{N_{\mathrm{TM}}} 10^{\frac{L_{\mathrm{TM}}\left[z(j),z(i)\right]}{10}} + \sum_{j=1}^{N_{\mathrm{NM}}} 10^{\frac{L_{\mathrm{NM}}\left[z(j),z(i)\right]}{10}}\right],$$

$$(2.23)$$

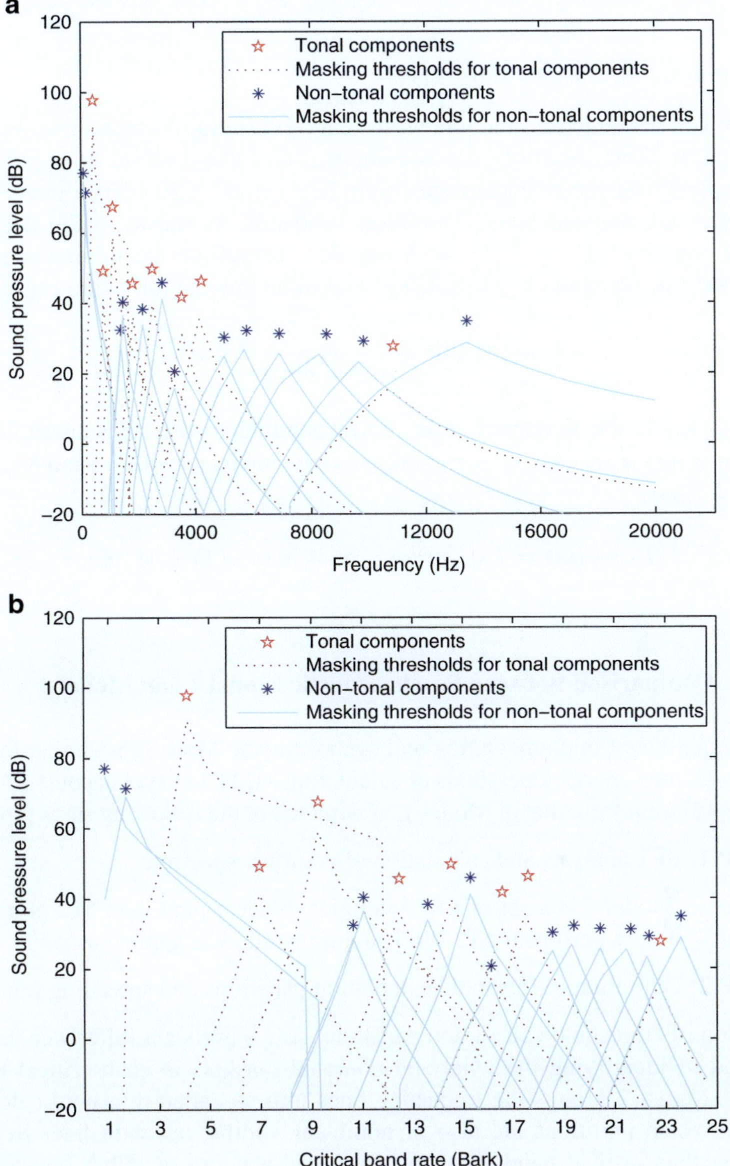

Fig. 2.21 Individual masking thresholds (**a**) Frequency on linear scale (**b**) Frequency on Bark scale

where ATH (i) is the SPL of threshold in quiet at frequency index i, N_{TM} and N_{NM} are the number of tonal and nontonal maskers, and $L_{\text{TM}}\left[z\left(j\right),:\right]$ and $L_{\text{NM}}\left[z\left(j\right),:\right]$ are their corresponding individual masking thresholds.

The global masking threshold is denoted by a bold dashed black line in Fig. 2.22.

- **STEP 6:** Determination of the MMT

The MMT is derived from the global masking threshold. As mentioned in Step 4, the global masking threshold L_G is computed on only a subset of samples (here 106 samples) over the frequency spectrum, i.e., $1 \leq i \leq 106$. Then these spectral subsamples are mapped onto 32 uniform subbands, as shown in Fig. 2.23. Each subband contains $\frac{N/2}{32} = \frac{512/2}{32} = 8$ samples. Therefore, the minimum masking level in the nth subband ($1 \leq n \leq 32$) is determined by the following expression:

$$L_{\mathrm{Min}}(n)/\mathrm{dB} = \min_{f_{id}(i) \in \mathrm{subband}\, n} L_G(i), \tag{2.24}$$

where $f_{id}(i)$ is the frequency index corresponding to the ith subsample. After spreading every $L_{\mathrm{Min}}(n)$ ($1 \leq n \leq 32$) over its subband with 8 samples, we get the MMT L_{MMT}:

$$L_{\mathrm{MMT}}(m) = L_{\mathrm{Min}}(n) \qquad m = [8(n-1)+1] : 8n. \tag{2.25}$$

2.4.1.3 Comparison Between Psychoacoustic Model 1 and Model 2

The general idea of implementation on Psychoacoustic Model 2 is similar to Model 1. However, the concrete operations of calculating MMT in Psychoacoustic Model 2 are quite different from that of Model 1, as depicted in the following steps [78,81]:

- **STEP 1:** FFT analysis and calculation of complex spectrum

The input to Model 2 is a set of 1,024 samples, twice longer than 512-point frame in Model 1. Before performing FFT, a Hanning window is applied as well.

- **STEP 2:** Definition of threshold calculation partitions and spreading function

The notion of "threshold calculation partitions" is a significant difference in Model 2. Instead of identifying the tonal and nontonal maskers in each critical band in Model 1, Model 2 groups the frequency lines into so-called threshold calculation partitions. Such partitions are also of nonlinear widths, but with finer frequency resolution than critical band. Each partition has a width of either one FFT line (at low frequencies) or 1/3 critical band (at high frequencies), whichever is wider [78]. According to this criterion, there are 57 partitions at a sampling frequency of 44.1 kHz by calculation. The result complies with Table D.3b in [77].

The spreading function in Model 2 is described by Eq. (2.7) and one specific spreading function is defined for each partition. Note that $10\log_{10} \mathrm{SF}(dz)$ in Eq. (2.7) is level-independent and thereby suitable for alleviating the computational burden of convolution in Step 4.

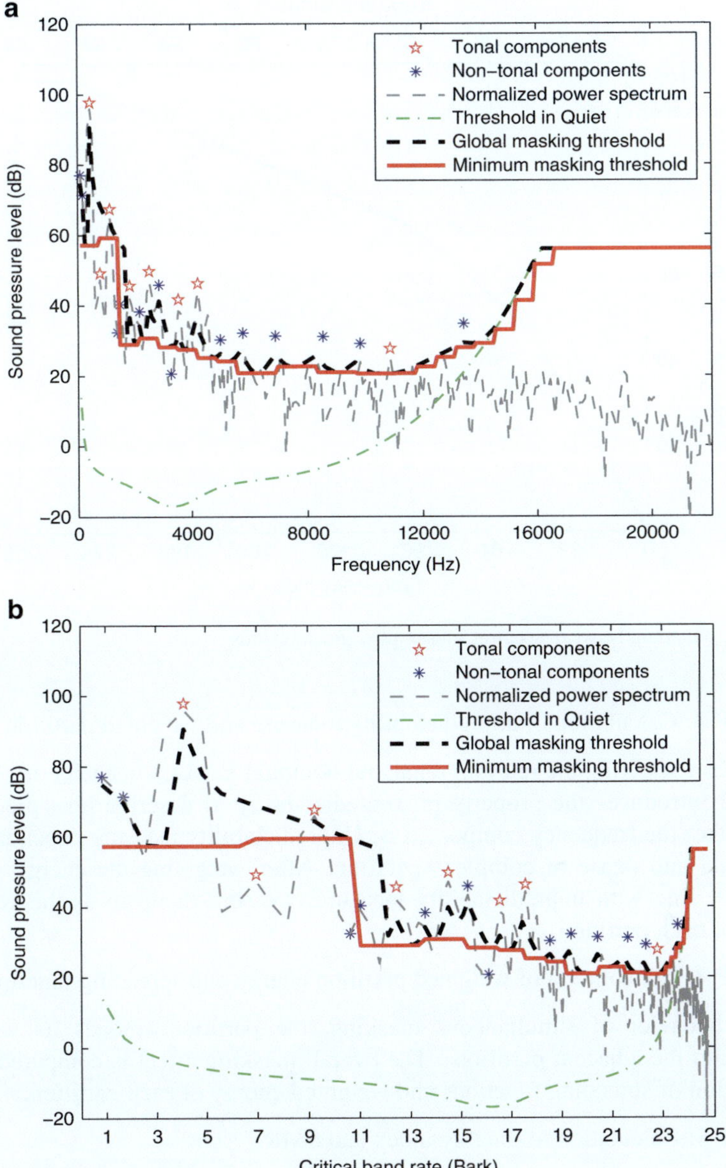

Fig. 2.22 Global masking threshold and minimum masking threshold (**a**) Frequency on linear scale (**b**) Frequency on Bark scale

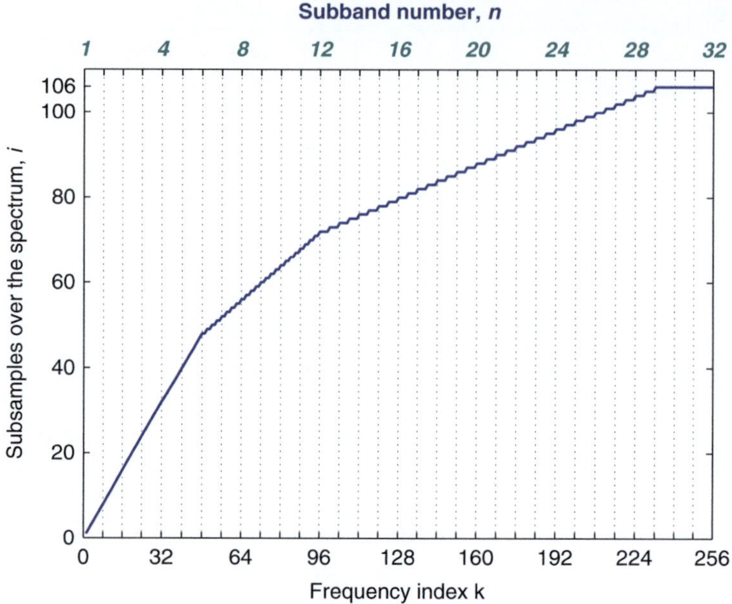

Fig. 2.23 Mapping between spectral subsamples and subbands

- **STEP 3:** Calculation of unpredictability measure and weighted partition energy

Rather than selecting the relevant tonal and nontonal maskers in each critical band, Model 2 introduces the property of unpredictability to describe how predictable (tonal -like) the frequency component is. Unpredictability measure depends on the magnitude and phase of complex spectrum. After weighting the energy of each frequency line with unpredictability measure, we sum them up as the weighted energy of each partition.

- **STEP 4:** Convolution of weighted partition energy and spreading function

As the behavior of simultaneous masking, the partition spreads its weighted energy into the adjacent partitions. The overall masking effect is computed by the convolution of spreading functions and weighted energy of each partition.

- **STEP 5:** Calculation of tonality index and SMR

Tonality index is a measure in Model 2, which is not used in Model 1. It denotes the relative tonality of the maskers in each partition. The value of tonality index is limited to the range of 0 (high unpredictability and noise-like) and 1 (low unpredictability and tonal). Based on tonality index as well as an attenuation shift factor between NMT and TMN, the SMR of each partition is calculated.

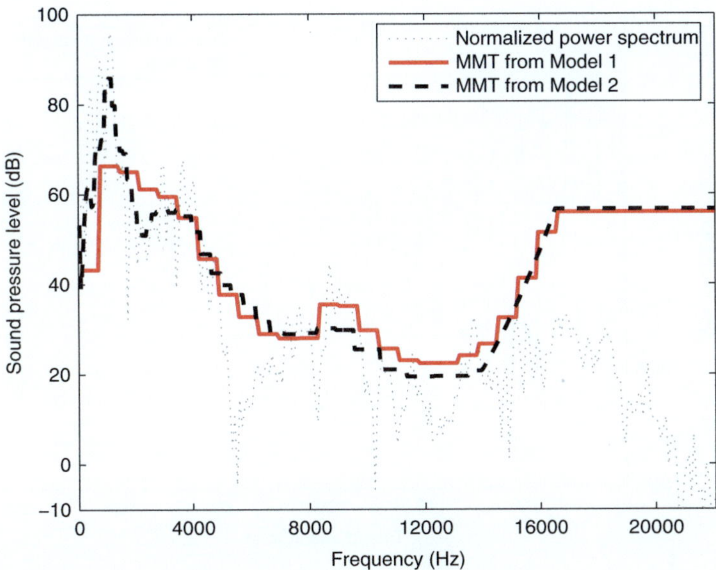

Fig. 2.24 Comparison of MMTs from Psychoacoustic Model 1 and 2

- **STEP 6:** Determination of the MMT

After obtaining SMR, the masking level of each partition is calculated by multiplying SMR to the inverse of signal energy, and then it spreads evenly over the frequency line(s) within the partition. Finally, the MMT is determined by taking the bigger value between the masking level and the threshold in quiet.

Figure 2.24 illustrates a comparison of the MMTs from Psychoacoustic Model 1 and 2. In view of the overall trend, MMT from Model 1 is analogous to that from Model 2, although a bit less accurate at low frequencies. Generally, the difference in masking effect of two psychoacoustic models is not evident [78]. On the other hand, as the price of high precision, Model 2 involves more calculations such as finer resolution of partitions, unpredictability measure, and the convolution process. Consequently, it slows down the speed of execution, which is against the requirement of audio watermarking. Therefore, we prefer Psychoacoustic Model 1 for our application.

2.4.2 Modelling the Effect of Nonsimultaneous Masking

In addition to simultaneous masking, the effect of nonsimultaneous masking is also well exploited for developing perceptual models.

In [51], a time-sliding window is adopted in modelling the effect of nonsimultaneous masking. To resemble pre- and post-masking curves in Fig. 2.10, a weighting

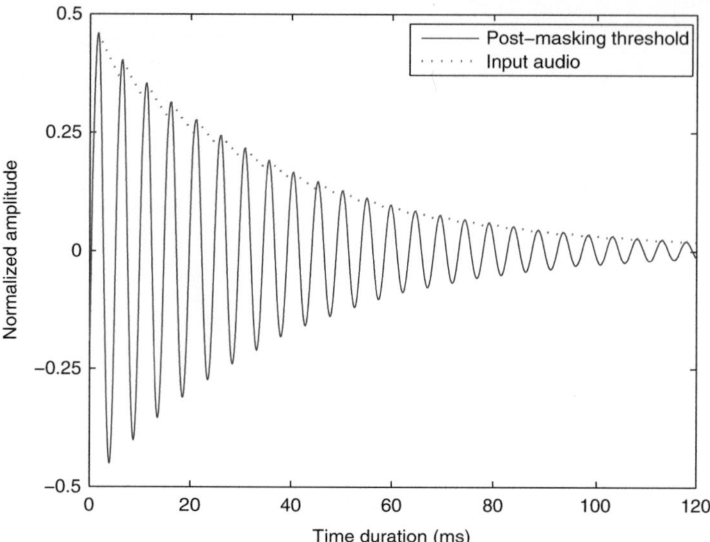

Fig. 2.25 Modelling the effect of post-masking

function of time is designed to be in a shape of bulge: a larger weight on components near the center of window, but gradual attenuation on components near the edges. Generally, it is assumed that such temporal smoothing is applied to signal spectrum, resulting in a smoothed output signal in time domain.

Different from [51], the modified envelope of input audio was used to approximate the effect of post-masking in [71]. In particular, the estimated masking curve increases with the envelope of signal and decays as an exponential function $e^{-\alpha t}$. The decay constant α ($\alpha \geq 0$) controls decaying rate as required, where $\alpha = 1.2 \times 10^{-3}$ in Fig. 2.25.

2.5 Summary

The ultimate aim of this chapter is to establish a psychoacoustic model that emulates the HAS. Accordingly, audio watermarking techniques are able to analyze the host audio signal in order to determine how the watermarks can be rendered as inaudible as possible.

The chapter started with the physiology of the peripheral auditory system including the outer, middle, and inner ears. The outer ear collects sound waves in the air and channels them to interior parts of the ear; the middle ear transforms the acoustical vibration of sound waves into mechanical vibration and passes them onto the inner ear; the inner ear transduces mechanical energy into nerve impulses that are transmitted to the brain. Then, some fundamental concepts of psychoacoustics such as SPL, loudness, human hearing range, threshold in quiet, and critical bandwidth were introduced. The notions of two types of auditory masking, i.e., simultaneous and nonsimultaneous masking, were also explained. In simultaneous masking, it is noted that the masking ability of narrowband noise is superior to pure tone. Based on the acquired knowledge, the ways of constructing the models for simultaneous and nonsimultaneous masking effects are investigated respectively, particularly simultaneous masking. After reviewing several models for the spreading of masking, we described the details of implementing Psychoacoustic Model 1 in ISO/MPEG standard, followed by a comparison with Model 2. On balance, two psychoacoustic models have similar perceptual quality, but Model 2 requires more computation than Model 1. Consequently, we adopted Psychoacoustic Model 1 in the audio watermarking scheme we developed in this book.

Chapter 3
Audio Watermarking Techniques

In recent years, there has been considerable interest in the development of audio watermarking techniques. To clarify the essential principles underlying a diversity of sophisticated algorithms, this chapter gives an overview of basic methods for audio watermarking, such as least significant bit (LSB) modification, phase coding, spread spectrum watermarking, cepstrum domain watermarking, wavelet domain watermarking, echo hiding, and histogram-based watermarking.

As the first step towards a full investigation into various approaches, we start with the details of performance evaluation undertaken in this book, including the parameters employed during perceptual quality assessment and robustness tests. Then, different audio watermarking techniques are separately implemented and evaluated in order to ascertain their advantages and disadvantages. Also, possible enhancements are exploited to further improve their capabilities. Finally, the chapter is concluded with a summary of comparative study.

3.1 Specifications on Performance Evaluation

As discussed in Sect. 1.3.2, performance evaluation of audio watermarking systems involves three major aspects, i.e., perceptual quality assessment, robustness test, and security analysis. To ensure a fair comparison of the different techniques, this section states the details of evaluation methods and test parameters. Without further notice, these specifications apply to all the performance evaluations hereafter, including the experiments carried out in this chapter and Chap. 5.

Y. Lin and W.H. Abdulla, *Audio Watermark: A Comprehensive Foundation Using MATLAB*,   51
DOI 10.1007/978-3-319-07974-5_3, © Springer International Publishing Switzerland 2015

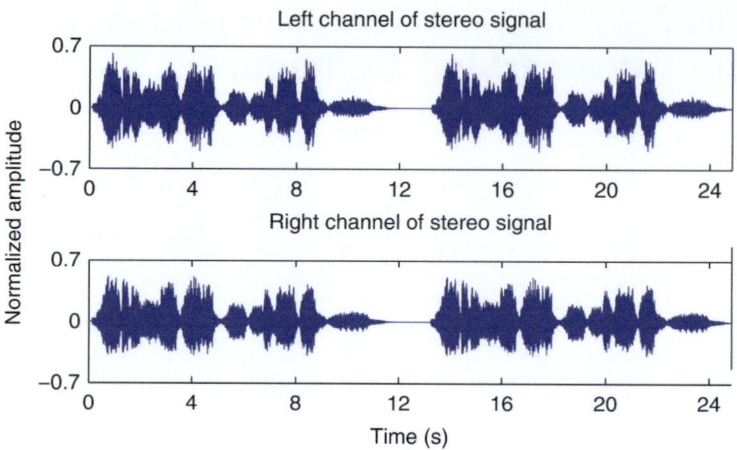

Fig. 3.1 An example of a two-channel stereo signal

3.1.1 Audio Test Signals Used for Evaluation

A collection of audio test files are prepared for performance evaluations in this book. The test set contains seventeen pieces of audio signals in total and all of them are in WAVE format (44.1 kHz, 16 bit, mono). For simplicity of reference, each audio signal A_n is marked with a subscript number n, i.e., (i) Vocal: Soprano$_1$, Bass$_2$, Quartet$_3$; (ii) Percussive instruments: Hihat$_4$, Castanets$_5$, Glockenspiel1$_6$, Glockenspiel2$_7$; (iii) Tonal instruments: Harpsichord$_8$, Violoncello$_9$, Horn$_{10}$, Pipes$_{11}$, Trumpt$_{12}$, Electronic tune$_{13}$; (iv) Music: Bach$_{14}$, Pop$_{15}$, Rock$_{16}$, Jazz$_{17}$. More details of each audio test file, such as its duration and waveform, are listed in Appendix E.

Except for the category of music, $A_{14} \sim A_{17}$, the majority of test files are selected from audio tracks on the EBU SQAM[1] disc, specifically for the testing and evaluation of sound systems [79]. Originally, most audio tracks are stereo channels. Since our study is not limited to stereo audio watermarking,[2] the left channel of each signal is always used for the watermarking. Figure 3.1 shows an example of a stereo signal and the left channel signal is taken as the audio test file A_2. It is observed that there are silent intervals inherent in audio data. Long silence is usually intractable in the watermarking, because embedding watermarks in muteness would unavoidably introduce perceived noise. Thus, watermarking

---

[1] EBU: The European Broadcasting Union; SQAM: Sound Quality Assessment Material.

[2] In general, stereo audio watermarking depends on some kind of relation between two channels [80], so it can only apply to stereo signals. However, mono audio watermarking can commonly treat one stereo channel as two mono channels, so it supports both mono and stereo audio signals.

regions must be carefully chosen. This issue will be further elaborated in Chap. 4. To avoid long silence during the watermarking, we only embed the watermark into the first half of A_2 for performance evaluations in this chapter.

3.1.2 Implementation of Perceptual Quality Assessment

As mentioned in Sect. 1.3.2.1, perceptual quality assessment on audio watermarking is comprised of subjective listening tests and objective evaluation tests.

Subjective listening tests are carried out in an isolated chamber, where ten trained listeners with different audio engineering backgrounds are participants. All the stimuli are presented through a high-fidelity headphone. In the MUSHRA test, the participants are asked to rate the perceptual quality of each watermarked signal relative to its host signal using a MATLAB's graphical user interface (GUI). The details of the developed GUI will be described in Sect. 5.2.1. Also, a rating based on the five-scale subjective difference grade (SDG) is performed as well. According to the descriptions in Table 1.2, the watermarked audio signals are expected to possess SDGs between -1.0 and 0. It is also acceptable if SDGs are less than -2.0 [2].

Moreover, software PEAQ [48] is utilized to provide an objective difference grade (ODG). ODG is an objective measurement of SDG and its specifications conform to those of SDG that are described in Table 1.2. The reason for choosing PEAQ rather than EAQUAL and PEMO-Q is that PEAQ is an improved version of EAQUAL and free to use, while PEMO-Q is a commercial software tool and its demo version is restricted to signal lengths up to 4 sec only. In addition, the signal-to-noise ratio (SNR) defined in Eq. (1.3) is calculated as an objective indicator of perceptual quality.

3.1.3 Implementation of Robustness Test

The general guideline for robustness tests was discussed in Sect. 1.3.2.2. On consideration of test items in SDMI, STEP2000, and StirMark for Audio, two robustness tests are set up in the book, i.e., a basic robustness test and an advanced robustness test.

Recall that in robustness tests, the attacked watermarked signals should not be degraded far beyond tolerable levels. On the basis of this premise, the attack parameters listed below are determined accordingly.

3.1.3.1 Basic Robustness Test

The basic robustness test incorporates a variety of typical attacks on audio watermarking techniques. Table F.1 in Appendix F shows the parameters used, expression, and implementation of each attack. The basic robustness test is employed for evaluating different audio watermarking techniques in Sect. 3.2 below. It is worth mentioning that the software "Adobe Audition v3.0"[3] is used to implement some attacks.

Noise addition: Add white Gaussian noise to the watermarked audio signal to reach the specified SNR. Usually, the SNR is targeted at a value between 20 and 40 dB.

Resampling: The watermarked audio signal that originally has a sampling rate of 44.1 kHz is downsampled to 22.05 or 11.025 kHz and then upsampled back to 44.1 kHz.

Requantization: The watermarked audio signal that originally has 16 bits/sample is requantized down to 8 bits/sample and then requantized back to 16 bits/sample.

Amplitude scaling: The amplitude of the watermarked audio signal is rescaled by $\pm 10\,\%$ or $\pm 20\,\%$. A positive and negative rate of scaling denotes that the amplitude is amplified and attenuated, respectively.

Low-pass filtering: A low-pass filter with a cutoff frequency of 4, 6, or 8 kHz is applied to the watermarked audio signal.

Echo addition: An echo with a delay of 100 or 200 ms and a decay of 20 % or 30 % is added to the watermarked audio signal.

Reverberation: Reverberation in a large empty hall with a reverberation time[4] of 1 s is exerted on the watermarked audio signal.

MP3 compression: The watermarked audio signal originally in .wav format is compressed at a bitrate of 48, 64, 96, or 128 kbps (kilobits per second) by an MP3 encoder. Then the .mp3 file is decompressed back to .wav format by the MP3 decoder. One point to note is that the process of compression/decompression causes not only amplitude modification but also displacement between the watermarked and attacked audio signals. A certain amount of quasi-zero samples are padded at the inception[5] and the end of attacked signal [81], because of internal data organization in MP3 files. Therefore, MP3 compression has two forms, i.e., Compression I and Compression II. In Compression I, we cut off those extra samples so as to focus on the effect of amplitude modification by MP3 compression. Meanwhile, Compression II is actually a combined attack that combines data compression and zeros inserting.

---

[3]Adobe Audition is a powerful digital audio recorder, editor, and mixer for Windows. It can perform a lot of operations, such as resampling, requantization, amplitude scaling, reverberation, MPEG compression, time stretching, and pitch shifting, on various formats of audio files, .au, .voc, .vox, .wav, and so on.

[4]The reverberation time of a room is the time that it takes for sound to decay by a certain level α dB once the source of sound has stopped [30]. T_{60} is when $\alpha = 60$ dB.

[5]Based on extensive operations with Adobe Audition v3.0, it is found that an amount of 1,201 samples is added to the beginning of an audio file.

DA/AD conversion: The watermarked audio signal is played through the audio player in a computer. Then the playback signal is recorded by connecting the headphone jack to the line-in jack on the sound card of the computer.

Random samples cropping: A number of 25 ms intervals are cropped at randomly selected positions in the front, middle, and rear of the watermarked audio signal.

Jittering: Jittering is an evenly performed form of random samples cropping. For our watermarked audio signal, 0.1–0.2 ms out of every 20 ms is cropped.

Zeros inserting: A number of 25 ms silent intervals are inserted into randomly selected positions in the front, middle, and rear of the watermarked audio signal.

Pitch-invariant time-scale modification (PITSM): The time-scale of the watermarked audio signal is stretched from $\pm 4\,\%$ up to $\pm 10\,\%$, whereas the audio pitch is preserved. Positive PITSM results in a longer duration with a slower tempo, while negative PITSM results in a shorter duration with a faster tempo.

Tempo-preserved pitch-scale modification (TPPSM): The pitch-scale of the watermarked audio signal is shifted from $\pm 4\,\%$ up to $\pm 10\,\%$, whereas the audio tempo is preserved. Positive TPPSM results in a higher pitch, while negative TPPSM results in a lower pitch.

The last five attacks belong to desynchronization attacks, which cause displacement between the encoder and decoder. Therefore, it is difficult to retrieve a watermark suffering from such hazardous attacks, especially PITSM and TPPSM.

3.1.3.2 Advanced Robustness Test

The advanced robustness test involves more stringent attacks than the basic robustness test and is specifically designed for rigorously evaluating our proposed audio watermarking algorithm to be described in Chap. 4. It consists of three parts: a test with StirMark for Audio, a test under collusion, and a test under multiple watermarking.

Collusion: We separately embed n different watermarks $w_o^{(1)}$, $w_o^{(2)}, \ldots, w_o^{(n)}$ into a host signal s_o and obtain n watermarked signals $s_w^{(1)}$, $s_w^{(2)}, \ldots, s_w^{(n)}$ correspondingly. Without loss of generality, these watermarked signals are further combined to create n average watermarked signals $s_w^{\overline{(i)}}$ $(1 \leq i \leq n)$ as follows:

$$\begin{cases} s_w^{(j)} = \text{Embedding}\left(s_o, w_o^{(j)}\right), & 1 \leq j \leq n, \\ s_w^{\overline{(i)}} = \frac{1}{i}\left[s_w^{(1)} + s_w^{(2)} + \cdots + s_w^{(i)}\right], & 1 \leq i \leq n. \end{cases} \tag{3.1}$$

In the detection, i watermarks $w_e^{(i,\cdot)}$ are detected from the average watermarked signal $s_w^{\overline{(i)}}$ individually:

$$w_e^{(i,j)} = \text{Detection}\left(s_w^{\overline{(i)}}\right), \quad 1 \leq i \leq n \text{ and } 1 \leq j \leq i, \tag{3.2}$$

where $n = 2 \sim 4$ in our robustness test.

Multiple watermarking: n different watermarks $w_o^{(1)}, w_o^{(2)}, \ldots, w_o^{(n)}$ are sequentially embedded in the following way:

$$\begin{cases} s_w^{(1)} = \text{Embedding}\left(s_o, w_o^{(1)}\right), \\ s_w^{(i)} = \text{Embedding}\left(s_w^{(i-1)}, w_o^{(i)}\right), & 2 \leq i \leq n. \end{cases} \tag{3.3}$$

In the detection, i watermarks are detected from the watermarked signal $s_w^{(i)}$ individually:

$$w_e^{(i,j)} = \text{Detection}\left(s_w^{(i)}\right), \quad 1 \leq i \leq n \text{ and } 1 \leq j \leq i, \tag{3.4}$$

where $n = 2 \sim 4$ in our robustness test.

In multiple watermarking, one point to note is that the technique for each watermarking may be the same or different. For example, the first watermark is embedded using our proposed algorithm, but the second watermark may be embedded using echo hiding watermarking. In this case, the corresponding detection method is employed to detect each watermark.

3.2 Audio Watermarking Algorithms

Over the last years, many digital watermarking methods have been proposed for different applications. These methods can be broadly divided into two main categories: (1) blind embedding, where the encoder does not exploit the knowledge of the host signal, for example, spread spectrum watermarking, and (2) informed embedding, where the knowledge of the host signal is adequately exploited by the encoder, for example, quantization index modulation (QIM) [1, 82].[6] Both prototypes have found implementations in audio watermarking, such as LSB modification, phase coding, spread spectrum watermarking, cepstrum domain watermarking, wavelet domain watermarking, echo hiding, and histogram-based watermarking. In the following subsections, one algorithm of each technique is implemented and evaluated separately.

Regarding the performance evaluations in this section, perceptual quality assessment and robustness tests are conducted as follows. In the perceptual quality assessment, the ODG using software PEAQ and the SNR are employed to indicate the audio quality. Since subjective listening tests are costly and time-consuming, only informal subjective listening tests are carried out if necessary. Informal subjective listening tests are performed in the same environment as described in Sect. 3.1.2; however, only a couple of listeners are involved. Without providing SDG scores, they are merely required to ascertain whether the watermarked signal is perceptually undistinguished from the host signal.

---

[6]Two categories are named as the host-interference nonrejecting method and the host-interference rejecting method respectively in [35, 83, 84].

For the robustness test, some or all the attacks in basic robustness test are involved and the bit error rates (BERs) as defined in Eq. (1.4) are calculated accordingly. Moreover, repetition coding on the watermark is adopted to enhance the robustness. For a $(n_r, 1)$ repetition code, each watermark bit is repeated n_r times and subsequently embedded. In the detection, the bits are determined using the majority vote rule. For example, repetition coding with $n_r = 3$ on the original watermark $w_o = \begin{bmatrix} 1 & 0 & 1 \end{bmatrix}$ yields a sequence $\hat{w}_o = \begin{bmatrix} 111 & 000 & 111 \end{bmatrix}$. Suppose that the attacked sequence becomes $\hat{w}_e = \begin{bmatrix} 011 & 001 & 110 \end{bmatrix}$, then the final detected watermark is $w_e = \begin{bmatrix} 1 & 0 & 1 \end{bmatrix}$ and BER = 0 %.

3.2.1   Least Significant Bit Modification

Earlier audio watermarking techniques embed the watermark into the host signal in a straightforward manner. One method is to replace the LSB of each sample with the watermark represented in a coded binary string [36]. In this way, the data payload of the watermarking system could be very high, approximately of the same order of magnitude with the sampling frequency of the host signal. Ideally, for instance, the bit rate is 44.1 kbps for an audio signal with a sampling frequency of 44.1 kHz [31]. However, the system under such conditions would be quite irresistible to any attack. In order to enhance the robustness and security, LSB modification can be performed on some selected subsets of the samples only, such as low-frequency components that are perceptually important. Usually, repetition coding could help increase the detection rate in LSB watermarking.

3.2.1.1   Algorithm

The audio watermarking technique in [85] is an example of the improved LSB modification, developed on the basis of the method in [86]. A host signal is first decomposed by a L-level complementary filterbank. Then the output from the low-pass filter is scrambled by a pseudorandom sequence (PRS) to increase the security. After applying the modified discrete cosine transform (MDCT), the coefficients at different orders are quantized to embed the watermark. In MATLAB, sound waves are normalized to a magnitude of one. Therefore, the coefficients are rounded at different decimal levels in the implementation of [85]. Moreover, repetition coding with $n_r = 4$ is used to reduce bit errors in the detection. Typically, it is a kind of LSB modification when the last decimal number is altered.

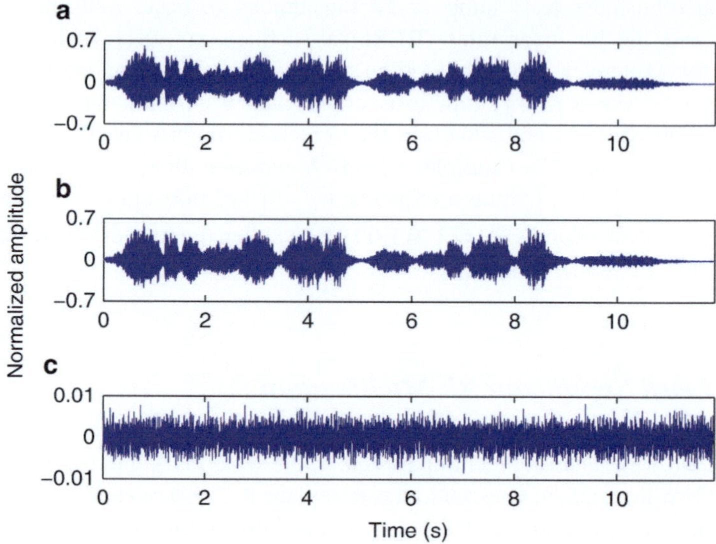

Fig. 3.2 Host signal and a watermarked signal by LSB modification. Note that the watermarked signal is produced by using $L = 6$ and modifying the third and fourth decimal places. (**a**) Host audio signal. (**b**) Watermarked audio signal. (**c**) Difference between the watermarked and host audio signals

3.2.1.2 Performance Evaluation

The performance of the improved LSB modification is evaluated in this section. Figure 3.2 shows the host signal and a watermarked signal by the improved LSB modification. Also, the results of performance evaluation are summarized in Table 3.1. As illustrated in Fig. 3.2c, the difference between the watermarked and host audio signals is small, which corresponds to a high SNR (i.e., 52.17 dB) in Table 3.1.[7]

From Table 3.1, the determination of decimal places results in a compromise between imperceptibility and robustness. On one hand, it is better to modify insignificant bits for the good of imperceptibility, such as the third or fourth decimal places. On the other hand, it is more robust if significant bits are used for embedding the watermark, such as the first or second decimal places. Moreover, the selection of a complementary filterbank level controls the balance between imperceptibility and data payload. By employing the complementary filterbank with less levels, we can embed more watermarking bits; nevertheless, such acquirements are at the expense of the degradation of the SNR.

---

[7]Note that two of the ODGs in Table 3.1 are slightly positive, i.e., 0.13 and 0.03. According to its definition, the ODG should normally be in the range $[--4, 0]$. However, if the distortion caused by watermarking is very low, then the cognitive model calculates positive values. In such cases, it is interpreted that the distortion is mostly inaudible for humans [38].

Table 3.1 Results of performance evaluation of LSB modification

Decimal place	Watermarking parameters							
	1st and 2nd		2nd and 3rd				3rd and 4th	
Filterbank level	8	10	4	6	8	10	4	6
Watermark length N_w	512	128	8192	2048	512	128	8192	2048
(1) Perceptual quality assessment								
SNR/dB	17.19	23.75	25.98	31.96	38.00	44.28	45.96	52.17
ODG	−3.18	−1.25	−3.01	−1.13	−0.32	0.13	−0.96	0.03
(2) Robustness test (BER: %)								
No attack	0	0	0	0	0	0	0	0
Noise (36 dB)	0	0	0.98	1.03	0.59	3.13	22.13	23.29
Lp filtering (8 kHz)	0.78	0.78	24.26	27.78	17.38	20.31	46.36	47.46
Echo (0.3, 100 ms)	13.28	15.63	42.44	47.27	41.41	42.97	48.56	49.71
Compression I (96 kbps)	0.39	0.78	10.88	13.43	11.33	14.06	43.69	45.17
Compression II (96 kbps)	46.68	39.06	49.29	48.83	49.61	40.63	49.74	50.73

On the whole, the watermarked signals produced by rounding off different
decimal places are robust against the attacks to some extent, except that the
detections fail completely under a desynchronization attack—Compression II.
However, LSB modification in the strict sense would survive only in closed, digital-
to-digital environments.

3.2.2 Phase Coding

Based on the fact that the human auditory system (HAS) is unable to perceive
the absolute phase, only the relative phase [36], audio watermarking techniques
can embed watermarks into the phase of host signal, i.e., phase coding and phase
modulation [31].

3.2.2.1 Algorithm

The basic phase-coding method was presented in [36]. It splits the host signal into
frames and the first frame's phase spectrum is modified to represent the watermark.
Then the phases of subsequent frames are changed accordingly to preserve their
relative phases. Thus, the first frame is crucial for watermark embedding and it
must not be an absolute silence. For watermark detection, the first frame of the
attacked signal is taken out, and then the value of its phase spectrum is calculated
to determine the watermark. The premise for detection is precise synchronization to
obtain the first frame accurately.

Algorithm 3.1 describes the pseudocode of phase coding. Note that the length of each frame is N, but the watermark has a length of $(N2 - 1)$ only, where $N2 = \lfloor \frac{N}{2} \rfloor$ and $\lfloor \cdot \rfloor$ is the smallest integer value. The reason is that the Fourier transform of a real-valued signal exhibits conjugate symmetry [87] and only half of the spectrum is available to embed the watermark. Moreover, the first spectral component is DC value whose phase is always equal to 0 or π, so it is not used for watermarking.

MATLAB script for phase-coding method can be found as *Phase − coding.m* file under Audio_Watermarking_Techniques folder in the attached CD.

3.2.2.2   Performance Evaluation

Table 3.2 shows the results of performance evaluation of the basic phase-coding method. As indicated by very low SNRs, the watermarked signals have been changed greatly and their perceptual quality has degraded badly.

The cause of deterioration is the substitution of Φ_1 with a binary sequence of $\frac{\pi}{2}$ and $-\frac{\pi}{2}$. In the basic method, every component of the phase spectrum is altered to represent the watermark. Such a sharp phase transition is likely to produce audible distortion [36]. In order to smooth the variation, the modified method is to change the phase spectrum at an interval of n_e and perform interpolation between the values. Several kinds of interpolation such as linear interpolation and cubic spline interpolation were tested.

Figure 3.3 illustrates an example of the watermarked signal produced by the modified phase-coding method. From Fig. 3.3c, there is still quite a difference between the watermarked and host signals, which also interprets very low SNRs in Table 3.2. Informal subjective listening tests show that the perceptual quality of the watermarked signals is not satisfied yet. Moreover, as shown in Table 3.2, the watermarked signals by phase coding are not robust and nearly all the BERs are over 20 %. In addition, repetition coding is not helpful to phase coding. This is because the effect of n_r times repetition coding resembles that of embedding interval $n_e = n_r$ with linear interpolation.

To achieve imperceptible watermarking, Kuo et al. [88] proposed phase modulation under the following constraint condition:

$$\left| \frac{d\phi(z)}{dz} \right| < 30°, \tag{3.5}$$

where $\phi(z)$ denotes the signal phase and z is the Bark scale [51]. More information on phase modulation for audio watermarking can be found in [31, 80, 81].

Algorithm 3.1 Pseudocode of basic phase coding method

% Watermark embedding
% Note: Host signal $s_o(n)$, $1 \leq n \leq N_o$ is split into N_P frames $\{g_i\}$ with N samples,
%   where $N_P = \lfloor \frac{N_o}{N} \rfloor$.

$s_w = [\,]$; % Initialize the watermarked signal $s_w(n)$
$s_o(n) = \{g_i(j)\}$, $1 \leq i \leq N_P$ and $1 \leq j \leq N$;
Binary watermark $w_o(m)$, $1 \leq m \leq (N2 - 1)$; % $N2 = \lfloor \frac{N}{2} \rfloor$

% Loop through every frame g_i
for $i = 1 : N_P$
  % Calculate the amplitude spectrum Γ_i and phase spectrum Φ_i
  $[\Gamma_i, \Phi_i] = F(g_i)$; % $F(\cdot)$ is the Fast Fourier Transform (FFT).
  if $i = 1$
    $\Psi_i(1) = \Phi_i(1)$;
    % Embed the watermark into the phase of the first frame
    for $m = 1 : (N2 - 1)$
      if $w_o(m) = 1$, then $\Psi_i(m + 1) = \frac{\pi}{2}$;
      if $w_o(m) = 0$, then $\Psi_i(m + 1) = -\frac{\pi}{2}$;
    end
  else
    $\triangle \Phi_{i-1} = \Phi_i(1 : N2) - \Phi_{i-1}(1 : N2)$;
    $\Psi_i = \Psi_{i-1} + \triangle \Phi_{i-1}$;
  end
  $\Omega_i^f = exp(j\Psi_i)$; % The first half of the phase spectrum
  % Function $fliplr(\cdot)$ is to flip a matrix left to right.
  % Function $conj(\cdot)$ is to compute the complex conjugate.
  $\Omega_i^s = fliplr\left(conj\left(\Omega_i^f\right)\right)$; % The second half of the spectrum
  $\Omega_i = \left[\Omega_i^f, 0, \Omega_i^s(1 : N2 - 1)\right]$;
  % Reconstruct the watermarked frame g_i^w
  $g_i^w = F^{-1}(\Gamma_i, \Omega_i)$; % $F^{-1}(\cdot)$ is the Inverse Fast Fourier Transform (IFFT).
  $s_w = \left[s_w, g_i^w\right]$.
end

% Watermark detection
% Note: s_a is the attacked signal.

$g_1^a = s_a(1 : N)$; % The first frame
$\left[\Gamma_1^a, \Phi_1^a\right] = F(g_1^a)$;
% The watermark w_e is detected based on the phase spectrum Φ_1^a.
for $m = 1 : (N2 - 1)$
  if $\Phi_1^a(m) \geq 0$, then $w_e(m) = 1$;
  if $\Phi_1^a(m) < 0$, then $w_e(m) = 0$.
end

Table 3.2 Results of performance evaluation of phase coding

Phase-coding method	Basic method			Modified method			
Frame length N	1,024	2,048	4,096	2,048	2,048	4,096	4,096
Embedding interval n_e	1	1	1	128	64	128	64
Watermark length N_w	511	1,023	2,047	8	16	16	32
(1) Perceptual quality assessment							
SNR/dB	−2.55	−2.53	−2.62	0.23	0.26	−2.15	−1.35
(2) Robustness test (BER: %)							
No attack	0	0	0	0	0	0	0
Noise (36 dB)	19.57	29.03	40.35	12.50	18.75	31.25	43.75
Lp filtering (8 kHz)	34.25	37.44	35.66	25.00	25.00	37.50	21.88
Compression I (96 kbps)	18.98	21.02	23.69	37.50	50.00	37.50	59.38
Compression II (96 kbps)	54.21	44.97	49.73	87.50	50.00	50.00	37.50

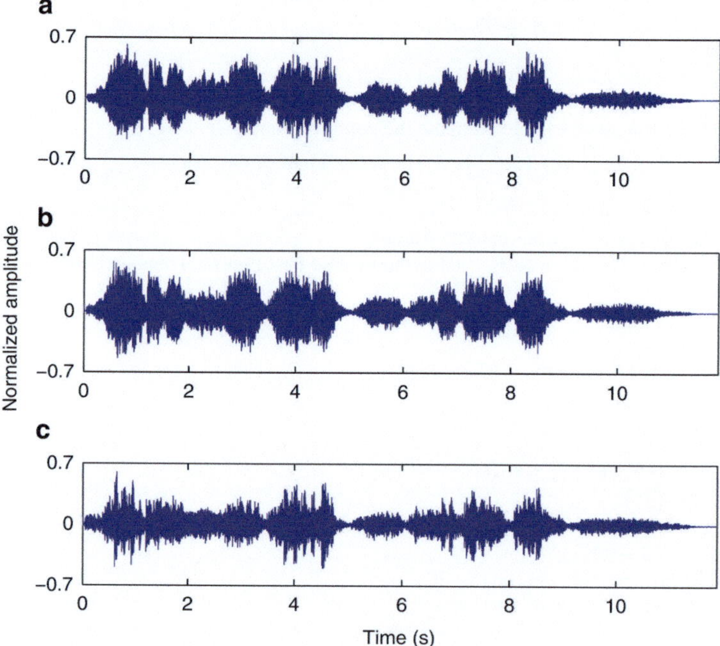

Fig. 3.3 Host signal and a watermarked signal by the modified phase-coding method. Note that the watermarked signal is produced by watermarking with $N = 2,048$ and $n_e = 128$. (**a**) Host audio signal. (**b**) Watermarked audio signal. (**c**) Difference between the watermarked and host audio signals

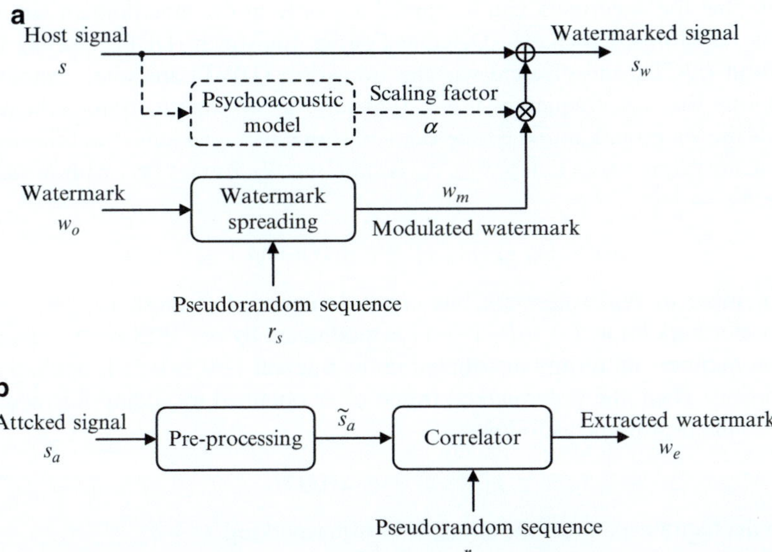

Fig. 3.4 Block diagram of basic SS watermarking scheme. (**a**) Embedding process. (**b**) Detection process

3.2.3   Spread Spectrum Watermarking

Spread spectrum (SS) watermarking is considered to be the most popular technique for digital watermarking [4, 31]. It spreads the watermark throughout the spectrum of the host signal, so that the signal energy present in every frequency bin is very small and hardly detectable. In this way, the embedded watermark can possess a large measure of security as well as robustness [89]. However, the process of watermarking may easily introduce perceivable distortion to audio files. So amplitude shaping by the masking threshold from the psychoacoustic model is often employed to keep the watermark inaudible [1, 4, 90].

3.2.3.1   Algorithm

There are two main forms of SS watermarking, namely direct sequence spread spectrum (DSSS) [8, 89, 91] and frequency hopping spread spectrum (FHSS) [92]. DSSS-based audio watermarking method is more commonly used and its basic scheme is shown in Fig. 3.4 [1, 90]. In watermark embedding process, the watermark w_o is modulated by PRS r_s to produce the modulated watermark w_m. To keep w_m inaudible, scaling factor α may be used to control the amplitude of w_m. Then the watermarked signal s_w is produced by adding w_m to the host signal s_o. In watermark detection process, the watermark w_e is extracted by correlating the received signal s_a with the PRS r_s used in the embedding.

Note that the watermark can be spread not only in the time domain but also in various transformed domains. Discrete Fourier transform (DFT), discrete cosine transform (DCT), and discrete wavelet transform (DWT) are some examples of transforms that are frequently used. Typically, a SS watermarking scheme that spreads the watermark into the time-domain signal is implemented as follows.

First, host signal $s_o(n)$, $1 \leq n \leq N_o$ is split into N_P frames $\{g_i\}$ with N samples, where $N_P = \lfloor \frac{N_o}{N} \rfloor$:

$$s_o(n) = \{g_i(j)\}, \quad 1 \leq i \leq N_P \text{ and } 1 \leq j \leq N. \tag{3.6}$$

So a number of N_P watermark bits can be embedded at most, i.e., $N_w = N_P$. Each watermark bit $w_o(i) \in \{+1, -1\}$ is modulated by one PRS r_s. A sequence of random numbers uniformly distributed in the interval $(-0.5, 0.5)$ is applied in our experiment. Then, the watermarked frame g_i^w is obtained by adding the modulated frame to the host frame as follows:

$$g_i^w = g_i + \alpha w_o(i) r_s, \tag{3.7}$$

where the factor α controls the strength of watermarking.

For a better perceptual quality, adaptive factor $\alpha = \beta \max(\mathrm{abs}(g_i))$ is adopted, where β is a scaling factor. Finally, all the watermarked frames $\{g_i^w\}$, $1 \leq i \leq N_w$ are concatenated to produce the watermarked signal s_w.

In the detection, watermark bits are determined by using a linear correlation between the watermarked signal and PRS. After splitting the watermarked signal $s_w(n)$ into frames $\{g_i^w\}$ in the same way as the embedding, the linear correlation $R_c(\cdot)$ between each frame g_i^w and r_s is calculated as

$$R_c(i) = \frac{1}{N} \sum_{j=1}^{N} g_i^w(j) \cdot r_s(j)$$

$$= \frac{1}{N} \sum_{j=1}^{N} [g_i(j) + \alpha w_o(i) r_s(j)] \cdot r_s(j)$$

$$= \frac{1}{N} \sum_{j=1}^{N} g_i(j) \cdot r_s(j) + \frac{1}{N} \sum_{j=1}^{N} \alpha w_o(i) [r_s(j)]^2. \tag{3.8}$$

Ideally, if the host frame g_i and the PRS r_s are independent, the first term in Eq. (3.8) is close to zero. Meanwhile, the second term has a large magnitude and its sign depends on the watermark bit $w_o(i)$. However, it is not always the condition that g_i and r_s are uncorrelated. In this case, the first term has similar or even larger magnitude than the second term, which would lead to incorrect detection. Thus, the watermarked signal must be preprocessed to reduce the effect of the host signal to the fullest extent [90]. To this end, different preprocessing methods are developed, such as linear predictive coding (LPC) filtering [93], cepstrum filtering [8], the Savitzky–Golay filtering [4], and decorrelation by subtracting the host signal [89] or adjacent frames [9].

For simplicity of implementation here, we assume the host signal is known at the decoder, i.e., a non-blind watermarking. Then, g_i is subtracted directly from g_i^w to eliminate the first item in Eq. (3.8). After that, each watermark bit $w_e(i)$ is determined based on the correlation value $R_c(i)$:

$$w_e(i) = \begin{cases} 1, & \text{if } R_c(i) \geq 0, \\ 0, & \text{if } R_c(i) < 0. \end{cases} \tag{3.9}$$

MATLAB script for SS watermarking can be found as $SS - watermarking.m$ file under Audio_Watermarking_Techniques folder in the attached CD.

3.2.3.2   Performance Evaluation

Figure 3.5 shows the host signal and a watermarked signal by SS watermarking. As a result of a small difference in Fig. 3.5c, the SNRs in Table 3.3 are higher than 30 dB. For perceptual quality assessment, however, the values of the ODG are pretty low. To ascertain the real perceptual quality, we carry out an informal subjective listening test on the watermarked signal with the highest ODG in Table 3.3, i.e., the shaded column. When the watermarked signal is played at a high volume, a constant hissing background noise is heard. This occurs as a result of the modulated PRS added to the host signal as white noise. Therefore, amplitude shaping by the psychoacoustic model is very important to produce an unperceived watermark.

In the robustness test, SS watermarking with repetition coding behaves differently. In descending order of resistance, the attacks are sorted as Compression I, noise addition, requantization, low-pass filtering, echo addition, resampling, and Compression II. Apparently, SS watermarking is quite vulnerable to desynchronization attacks. Thus, proper solutions for the synchronization problem are necessary to improve the detection rate.

3.2.4   Cepstrum Domain Watermarking

Cepstrum domain watermarking is performed to embed the watermark into cepstral coefficients. By definition, the complex cepstrum $\hat{x}(n)$ is the inverse Fourier transform of the complex logarithm of the Fourier transform of a signal $x(n)$. Mathematically, it is described as follows [94]:

$$X\left(e^{j\omega}\right) = \sum_{n=0}^{N-1} x(n)\, e^{-j\omega n}$$

$$\hat{X}\left(e^{j\omega}\right) = \log\left\{X\left(e^{j\omega}\right)\right\} = \log\left|X\left(e^{j\omega}\right)\right| + j\,\arg\left[X\left(e^{j\omega}\right)\right] \tag{3.10}$$

$$\hat{x}(n) = \frac{1}{2\pi}\int_{-\pi}^{\pi} \hat{X}\left(e^{j\omega}\right) e^{j\omega n}\, d\omega,$$

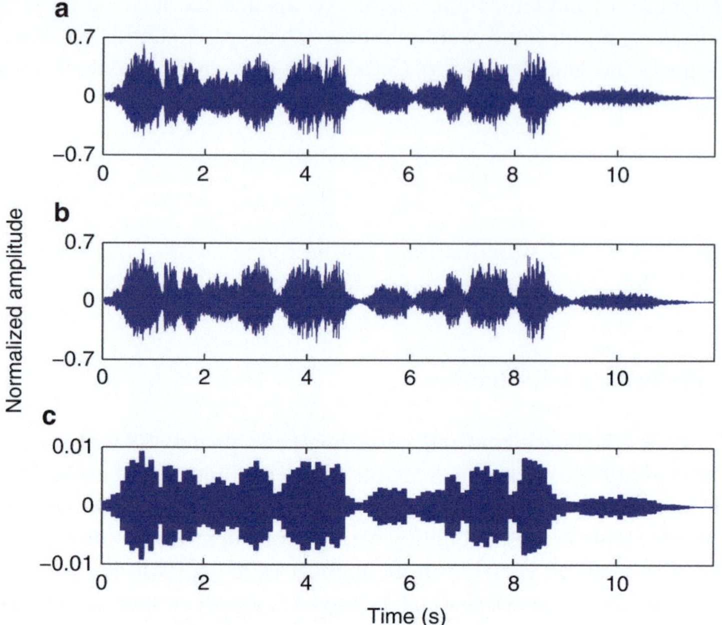

Fig. 3.5 Host signal and a watermarked signal by SS watermarking. Note that the watermarked signal is produced by watermarking with $N = 4,096$, $n_r = 3$ and $\beta = 0.03$. (**a**) Host audio signal. (**b**) Watermarked audio signal. (**c**) Difference between the watermarked and host audio signals

where $\log(\cdot)$ refers to the natural logarithm. In a more compact way, $\hat{x}(n) = F^{-1}\{\log(F\{x(n)\})\}$, where $F(\cdot)$ is the Fourier transform and $F^{-1}(\cdot)$ is its inverse.

The complex cepstrum $\hat{x}(n)$ has preserved information about the magnitude and phase of the frequency spectrum of $x(n)$. Therefore, $x(n)$ can be recovered from $\hat{x}(n)$ by the inverse complex cepstrum. Figure 3.6 shows the block diagram of computing the complex cepstrum and the inverse complex cepstrum using DFT and IDFT [87].

In MATLAB implementation, the real part of the complex cepstrum is often utilized as the common "cepstrum" $c(n)$ [95], i.e.,

$$c(n) = \text{real}(\hat{x}(n)). \tag{3.11}$$

Note that the real part of the complex cepstrum $c(n)$ should be distinguished from the "real cepstrum," $c_r(n)$. The real cepstrum is defined as the inverse Fourier transform of the logarithm of the magnitude of the Fourier transform of a signal $x(n)$, i.e.,

$$c_r(n) = \frac{1}{2\pi}\int\limits_{-\pi}^{\pi} \left|\hat{X}(e^{j\omega})\right| e^{j\omega n} d\omega. \tag{3.12}$$

Table 3.3 Results of performance evaluation of SS watermarking

	Watermarking parameters							
Frame length N	2048				4096			
Repetition coding n_r	1	3	1	3	1	3	1	3
Watermark length N_w	256	85	256	85	128	42	128	42
Scaling factor β	0.02	0.02	0.03	0.03	0.02	0.02	0.03	0.03
(1) Perceptual quality assessment								
SNR/dB	34.81	34.59	31.09	31.11	33.90	33.92	30.46	30.46
ODG	−3.68	−3.67	−3.76	−3.73	−3.70	−3.68	−3.77	−3.78
(2) Robustness test (BER: %)								
No attack	0	0	0	0	0	0	0	0
Noise (30 dB)	7.03	4.71	3.52	1.18	3.13	2.38	0.78	0
Re-sampling (22.05 kHz)	46.88	34.12	31.25	28.24	30.47	28.57	17.97	14.29
Requantization (8 bit)	4.69	5.88	2.34	0	4.69	0	1.56	0
Lp filtering (8 kHz)	12.89	11.76	4.69	0	7.03	0	0.78	0
Echo (0.3, 100 ms)	28.52	10.59	10.94	2.35	7.03	7.14	2.34	0
Compression I (96 kbps)	1.17	0	0.39	0	0	0	0	0
Compression II (96 kbps)	50.00	47.06	49.61	58.82	52.34	45.24	48.44	47.62

a

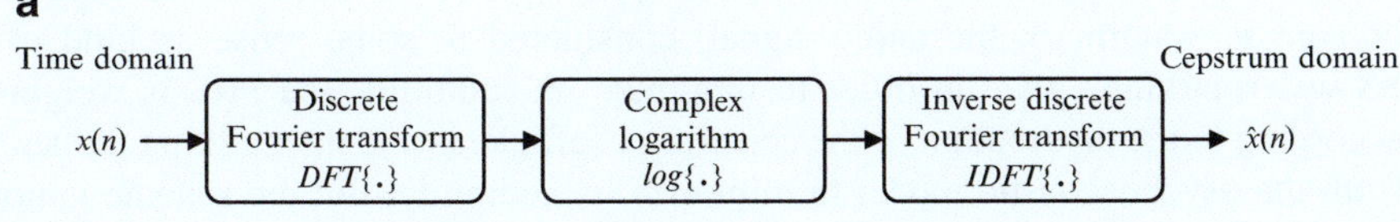

b

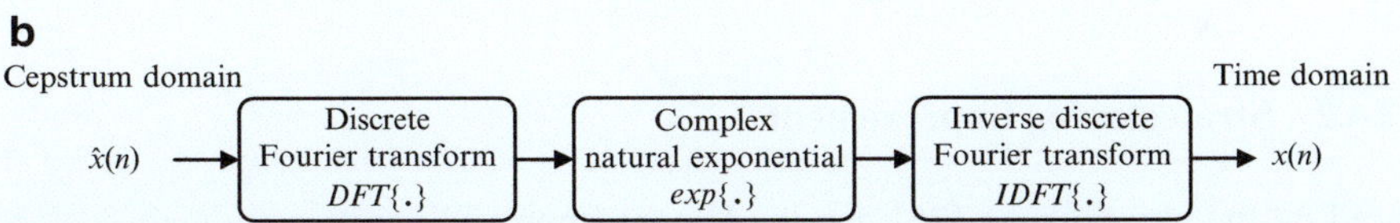

Fig. 3.6 Block diagram of computing the complex cepstrum and the inverse complex cepstrum. (**a**) Complex cepstrum $\hat{x}(n) = F^{-1}\{\log(F\{x(n)\})\}$. (**b**) Inverse complex cepstrum $x(n) = F^{-1}\{\exp(F\{\hat{x}(n)\})\}$

Different from complex cepstrum, real cepstrum has lost the phase information. Therefore, $x(n)$ cannot be reconstructed perfectly from $c_r(n)$.

3.2.4.1 Algorithm

There are some audio watermarking schemes conducting in the cepstrum domain [95–99]. Li et al. [95] found that the statistical mean of cepstrum coefficients is an attack-invariant feature and accordingly developed cepstrum domain watermarking based on statistical mean manipulation (SMM). In this way, the statistical mean of cepstrum coefficients, μ, is set to be $\pm\alpha_w$ that represent watermark bit "1" and "0" respectively. Apparently, the strength of watermarking is controlled by α_w. For the detection, the statistical mean of watermarked cepstrum coefficients μ^w is calculated and compared to a predefined threshold T_d. If $\mu^w \geq T_d$, then the watermark bit is "1." Otherwise, it is "0." Note that instead of the mean, we use the sum for bit determination in practice. This is because the mean of cepstrum coefficients is usually around 10^{-4}, which is rather small for a comparison.

Algorithm 3.2 describes the pseudocode of the watermarking process.

Furthermore, Hsieh et al. [97] proposed a method of embedding based on time energy features to solve the synchronization problem. The watermark is embedded into the frames followed by salient points, the positions where signal energy climbs fast to a peak [33]. As salient points are supposed to remain stable after attacks, synchronization can be regained in the detection. Afterward, Cui et al. [98] improved the method in [95] by employing the psychoacoustic model to control the audibility of the introduced distortion. Apart from SMM, Gopalan [99] embedded the watermark by altering the cepstrum in the regions that are psychoacoustically masked, so as to ensure a better trade-off between imperceptibility and robustness.

Different from [95, 97–99], Lee et al. [96] spread the PRS in the cepstrum domain to watermark the audio signal, considered in some sense as kind of a SS watermarking. Also, in order to minimize its audibility, the PRS is weighted according to the distribution of the cepstrum coefficients and the masking threshold from the psychoacoustic model to minimize its audibility. But the scheme is non-blind because the host signal is required in the detection.

3.2.4.2 Strategies for Improvement

Based on the algorithm in [95], we implement improved cepstrum domain watermarking. Several strategies are used to enhance the system performance. First of all, [95] did not mention the smooth transition between adjacent frames. As a result, continuous clicking sounds are clearly evident when playing the watermarked audio signal. To get rid of such noise, Hanning windowing and half overlapping (i.e., overlap factor $p = \frac{1}{2}$) are utilized to smooth the edges.

Secondly, since SMM is a statistical method, repetition coding is expected to help increase the overall detection accuracy. This is because repetition coding can help maintain the statistical properties of successive frames. Therefore, n_r times repetition coding ($n_r = 3$ or 5) is employed in the simulations.

Thirdly, detection threshold T_d is estimated by performing a number of pre-attack experiments as follows. The watermarked signal is pre-attacked by some

---

Algorithm 3.2 Pseudocode of cepstrum domain watermarking

% Watermark embedding

% Note: Host signal $s_o(n)$, $1 \leq n \leq N_o$ is split into N_P frames $\{g_i\}$ with N samples,

%   where $N_P = \lfloor \frac{N_o}{N} \rfloor$. Also, $N_w = N_P$.

$s_w = [\,]$; % Initialize the watermarked signal $s_w(n)$

$s_o(n) = \{g_i(j)\}$, $1 \leq i \leq N_P$ and $1 \leq j \leq N$;

Binary watermark $w_o(i)$, $1 \leq i \leq N_w$;

% Loop through every frame g_i

for $i = 1 : N_P$

      % Function $cceps(\cdot)$ is to compute the real part of the complex cepstrum.

      $c_i = cceps(g_i)$;

      $\mu_i = mean(c_i)$

      if $w_o(i) = 1$, then $c_i^w = c_i - \mu_i + \alpha_w$;

      if $w_o(i) = 0$, then $c_i^w = c_i - \mu_i - \alpha_w$;

      % Function $icceps(\cdot)$ is to compute the inverse complex cepstrum.

      $g_i^w = icceps\left(c_i^w\right)$;

      $s_w = \left[s_w,\, g_i^w\right]$

end

% Watermark detection

% Notes: $s_a(n)$ is the attacked signal.

$s_a(n) = \left\{g_i^a(j)\right\}$, $1 \leq i \leq N_P$ and $1 \leq j \leq N$;

Predefined threshold T_d;

% Loop through every frame g_i^a

for $i = 1 : N_P$

      $c_i^a = cceps\left(g_i^a\right)$;

      $\xi_i = sum\left(c_i^a\right)$;

      if $\xi_i \geq T_d$, then $w_e(i) = 1$;

      if $\xi_i < T_d$, then $w_e(i) = 0$.

end

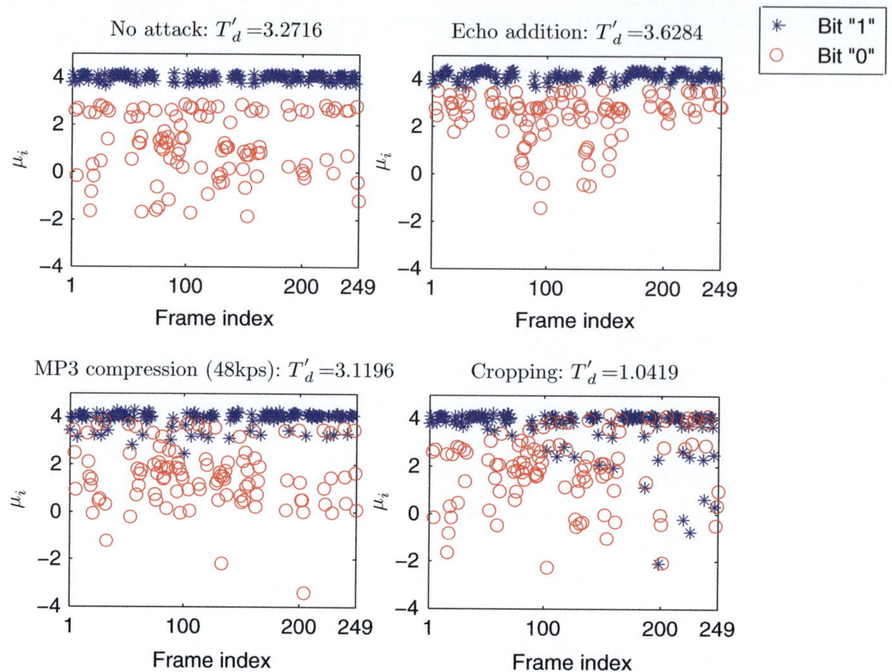

Fig. 3.7 Distributions of R_{one} and R_{zero} under different attacks. Note that these data are produced by watermarking with $N = 4,096$, $\alpha_w = 0.001$, and $n_r = 3$

commonly used attacks, such as noise addition, low-pass filtering, echo addition, MP3 compression, and random samples cropping. For every attacked signal, the sum of the cepstrum coefficients of each frame is written as $\mu_i = \text{sum}\left(c_i^a\right)$. If bit "1" was originally embedded in the ith frame, μ_i is put into the "one" set, i.e., $R_{\text{one}} = \{\mu_i \mid w_o(i) = 1\}$. Otherwise, μ_i is put into the "zero" set, i.e., $R_{\text{zero}} = \{\mu_i \mid w_o(i) = 0\}$. Then, the minimum of R_{one} and the maximum of R_{zero} are averaged to be one sub-threshold $T_d' = \frac{\min(R_{\text{one}}) + \max(R_{\text{zero}})}{2}$. Experimental results show that the elements in R_{one} are positive numbers, commonly being around $N\alpha_w$, whereas the elements in R_{zero} vary greatly between $[-N\alpha_w, N\alpha_w]$. Therefore, the maximum sub-threshold is a proper T_d to achieve low BERs. To further reduce the BER, more delicate adjustments on T_d are required. For a better illustration, the distribution of elements in R_{one} and R_{zero} is plotted in Fig. 3.7, where they are denoted by blue asterisks and red circles respectively. As shown on the graph, the maximum sub-threshold $T_d' = 3.6284$ under echo addition is an appropriate T_d. Furthermore, after subtle alteration, $T_d = 3.5$ is utilized as the final detection threshold for the conditions of $N = 4,096$, $\alpha_w = 0.001$, and $n_r = 3$ in Table 3.4.

Finally, variable frame length in the detection is utilized to combat pitch-invariant time-scale modification (PITSM). Under the default setting for SMM detection, the attacked signal $s_a(n)$, $1 \leq n \leq N_a$ is assumed to be as long as the watermarked

Table 3.4 Results of performance evaluation of cepstrum domain watermarking

	Watermarking parameters								
Frame length N	2048						4096		
Watermarking strength α_w	0.001			0.0015			0.001		
Watermark length N_w	504	168	100	504	168	100	251	83	50
Repetition coding n_r	1	3	5	1	3	5	1	3	5
Detection threshold T_d	1.6			2.5			3.5		
(1) Perceptual quality assessment									
SNR/dB	27.07	26.49	26.73	20.74	20.28	20.06	17.92	17.37	17.37
ODG	0.16	0.17	0.16	−0.46	−0.44	−0.49	0.04	0.08	0.07
(2) Robustness test (BER: %)									
No attack	25.99	8.93	2.00	18.45	0	0	11.55	0	0
Noise (30 dB)	27.98	10.12	3.00	18.25	0	0	11.55	0	0
Resampling (22.05 kHz)	26.79	8.33	3.00	18.45	0	0	11.55	0	0
Requantization (8 bit)	26.59	9.52	2.00	18.25	0	0	11.55	0	0
Amplitude +10 %	27.18	8.93	4.00	18.45	0	0	11.55	0	0
-10 %	27.78	10.12	5.00	19.44	0	0	12.35	0	0
Lp filtering (8 kHz)	26.39	8.33	2.00	18.25	0	0	13.15	0	0
DA/AD (line-in jack)	48.41	55.36	52.00	48.41	51.19	53.00	53.78	57.83	56.00
Echo (0.3, 200 ms)	27.58	2.38	4.00	20.83	0.60	0	13.94	0	0
Reverb (1 s)	28.17	12.50	4.00	18.85	0	0	14.34	0	0

(continued)

Table 3.4 (continued)

			Watermarking parameters								
Frame length N			2048						4096		
Watermarking strength α_w			0.001			0.0015			0.001		
Compression I	96 kbps		25.99	8.33	3.00	18.45	0	0	12.75	0	0
	64 kbps		26.79	10.71	3.00	17.86	0	0	12.35	0	0
	48 kbps		27.58	10.71	3.00	19.05	0	0	11.55	0	0
Cropping (4×25 ms)			45.83	35.71	22.00	42.66	29.76	24.00	35.46	16.87	0
Jittering (0.1/20 ms)			44.25	31.55	11.00	38.10	17.26	1.00	28.69	0	0
Inserting (4×25 ms)			49.21	39.29	24.00	42.06	29.76	23.00	31.47	18.07	0
PITSM	+10 %	N	49.40	45.83	45.00	46.43	47.02	46.00	52.19	48.19	44.00
		$\tilde{N}$	33.73	9.52	8.00	26.79	0.60	0	17.93	0	0
	−10 %	N	51.18	48.39	49.46	49.03	49.03	41.94	48.71	48.05	36.96
		$\tilde{N}$	34.72	16.07	13.00	24.6	0.60	0	15.14	0	0
TPPSM	+10 %		32.94	8.33	5.00	26.98	1.19	0	17.53	0	0
	−10 %		33.53	11.90	10.00	23.41	0.60	0	10.76	0	0

signal (as well as the host signal), i.e., $N_a = N_o$. Also, the detection uses the same frame length as the embedding, i.e., $\tilde{N} = N$. Then, the same splitting method adopted in the embedding is employed to divide $s_a(n)$ into $\tilde{N}_w = \left\lfloor \frac{N_a - p\tilde{N}}{\tilde{N}(1-p)} \right\rfloor = \left\lfloor \frac{N_o - pN}{N(1-p)} \right\rfloor = N_w$ frames. Correspondingly, N_w watermark bits are extracted.

However, PITSM will adjust playback speed of the audio signal and N_a is changed accordingly. For example, it is lengthened by positive PITSM ($N_a > N_o$) or shortened by negative PITSM ($N_a < N_o$). For most audio watermarking techniques, such alteration in signal length will cause severe problems to the detection. On one hand, both positive and negative PITSM modify the time-scale of the watermarked signal, which results in a displacement between the detection and embedding. Without the retrieval of synchronization, watermark detection cannot work properly. On the other hand, in the case of negative PITSM, we probably are unable to extract as many watermark bits as are embedded. This is because the attacked signal has not enough samples for the detection. For example, given $N_o = 5.252 \times 10^5$ and $N = 2,048$, the host signal is split into $N_w = \left\lfloor \frac{N_o - pN}{N(1-p)} \right\rfloor = \left\lfloor \frac{5.252 \times 10^5 - \frac{1}{2} \times 2,048}{2,048\left(1 - \frac{1}{2}\right)} \right\rfloor = 511$ frames, and hence a number of bits up to 511 can be embedded. After being modified by -10% PITSM, the attacked signal has a shorter length of $N_a = 4.77455 \times 10^5$. If $\tilde{N} = N = 2,048$ is still in use, there would be $\tilde{N}_w = \left\lfloor \frac{N_a - p\tilde{N}}{\tilde{N}(1-p)} \right\rfloor = \left\lfloor \frac{4.77455 \times 10^5 - \frac{1}{2} \times 2,048}{2,048\left(1 - \frac{1}{2}\right)} \right\rfloor = 465$ frames. This means that only 465 bits can be extracted at most,[8] not to mention the low detection accuracy.

Under such circumstances, variable frame length for re-synchronization in the detection is proposed: the frame length $\tilde{N}$ should vary with signal length N_a, i.e., $\tilde{N} = \left\lfloor N \cdot \frac{s_a}{s_o} \right\rfloor$. So the attacked signal would still be split into N_P frames approximately.

MATLAB script for cepstrum domain watermarking can be found as $Cepstrum - watermarking.m$ file under Audio_Watermarking_Techniques folder in the attached CD.

It is worth mentioning that although other desynchronization attacks such as random samples cropping, zeros inserting, and jittering may also change signal length, variable frame length is not very effective in these cases, especially the former two. Under serious cropping and inserting attacks, a large amount of removed or added samples occur locally, not uniformly along the whole watermarked signal. Therefore, the detection with variable frame length cannot achieve proper re-synchronization to recover the watermark. To withstand these attacks, either the watermark is embedded on the basis of attack-invariant features, such as the statistical mean of the cepstrum coefficients on a large scale, or the detection can locate the positions where the attacks take place, such as the synchronization method introduced in the next chapter.

---

[8]Despite the fact that only a portion of bits are extracted, the corresponding BER is always calculated for performance evaluations.

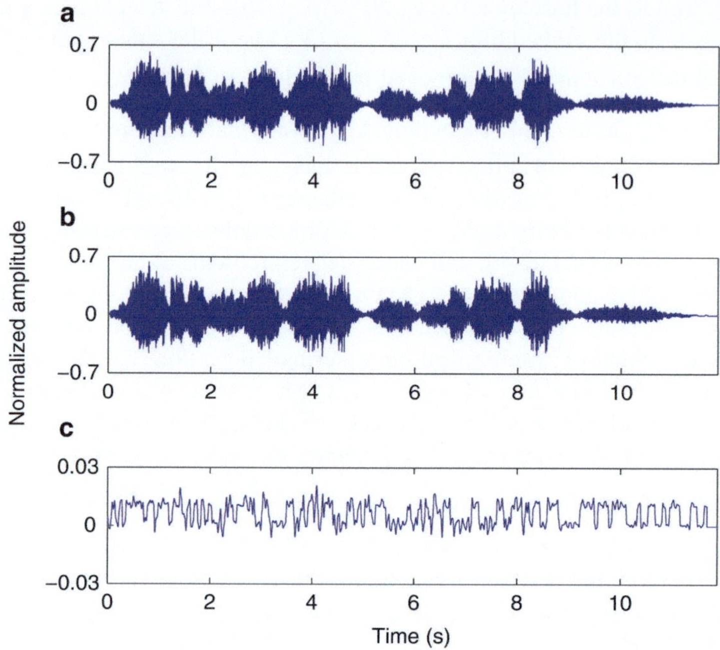

Fig. 3.8 Host signal and a watermarked signal by cepstrum domain watermarking. Note that the watermarked signal is produced by watermarking with $N = 2,048$, $\alpha_w = 0.0015$ and $n_r = 3$. (**a**) Host audio signal. (**b**) Watermarked audio signal. (**c**) Difference between the watermarked and host audio signals

3.2.4.3   Performance Evaluation

The watermarked signal illustrated in Fig. 3.8 is generated by improved cepstrum domain watermarking. Also, Table 3.4 shows the results of performance evaluation. Considering the trade-off between imperceptibility and robustness, watermarking strength α is set to be 0.001 or 0.0015. One point to note is that cropping and jittering would shorten the watermarked signal as well. In view of this, we usually do not embed watermark bits at full capacity, but slightly less bits. As calculated above, if $N = 2048$, the maximum capacity is $N_w = 511$ bits. In the experiments with $n_r = 1$, however, only 504 bits are embedded.

From Table 3.4, the overall performance of improved cepstrum domain watermarking is attractive. On one hand, the watermarked signals have pretty good perceptual quality in terms of high ODG scores. Informal subjective listening tests also show that the watermarked signals are perceptually undistinguished from the host signal.

On the other hand, with the help of repetition coding, the watermarked signals are quite robust against all the attacks except for DA/AD conversion. Since each frame's statistical property has been changed by half overlapping between neighboring frames, repetition coding is indispensable to watermark detection. For example, in the case of $N = 2,048$ and $\alpha_w = 0.001$, the watermarks without repetition coding ($n_r = 1$) have large BERs (more than 25 %). After triple repetition coding ($n_r = 3$), all the BERs excluding those under DA/AD conversion and some desynchronization attacks are dropped to 10 % or so. The BERs can be further reduced by quintuple repetition coding ($n_r = 5$). Also, a longer frame length would help increase the robustness as well. With the same watermarking strength and repetition coding, the BERs of the watermarks with $N = 4,096$ are much lower than those of $N = 2,048$. In addition, repetition coding and longer frame play an important role in improving the robustness against cropping and inserting, whereas watermarking strength is not really effective. For example, when $N = 4,096$ and $n_r = 5$, the watermarks attacked by cropping and inserting can be detected without any errors. This is because the statistical mean would keep stable on a large scale, which is required for correct detection of the watermark. However, loss of capacity is the price to pay for the robustness achieved by employing repetition coding and a longer frame.

Experimental results show that the detection with variable frame length (indicated as $\tilde{N}$) can successfully extract the watermarks from the watermarked signals attacked by PITSM. Moreover, the embedded watermarks are fairly resistant to TPPSM, where the BERs under ± 10 % TPPSM are still very low. This indicates that the statistical mean is immune to frequency variation caused by TPPSM.

On the whole, the improved cepstrum domain watermarking performs well in terms of imperceptibility and robustness. However, the issue of security is a challenge in real applications. Without prior knowledge, the attacker could deliberately make a slight modification to the mean of cepstrum coefficients of the watermarked signal, so that the watermark cannot be extracted properly.

3.2.5 Wavelet Domain Watermarking

In wavelet domain watermarking, the watermark is embedded into wavelet coefficients of host audio signal. Compared to Fourier transform, wavelet transform is more suitable to generate the time–frequency representation of nonstationary signals such as audio signals [100].

As shown in Fig. 3.9, DWT decomposition and reconstruction involve the multiresolution analysis and synthesis. During the decomposition, the input signal is first decomposed into two parts: the high-frequency part (i.e., the "detail" coefficients) from high-pass filter (H_0) and the low-frequency part (i.e., the "approximation" coefficients) from low-pass filter (G_0). Then, the low-frequency part is further decomposed into high- and low-frequency parts. This process is repeated until the desired level is reached. At each decomposition level, the time resolution is halved and the frequency resolution is doubled. In this way, DWT brings good frequency resolution at low frequencies and good time resolution at high frequencies. Since the human ear is more sensitive to low-frequency

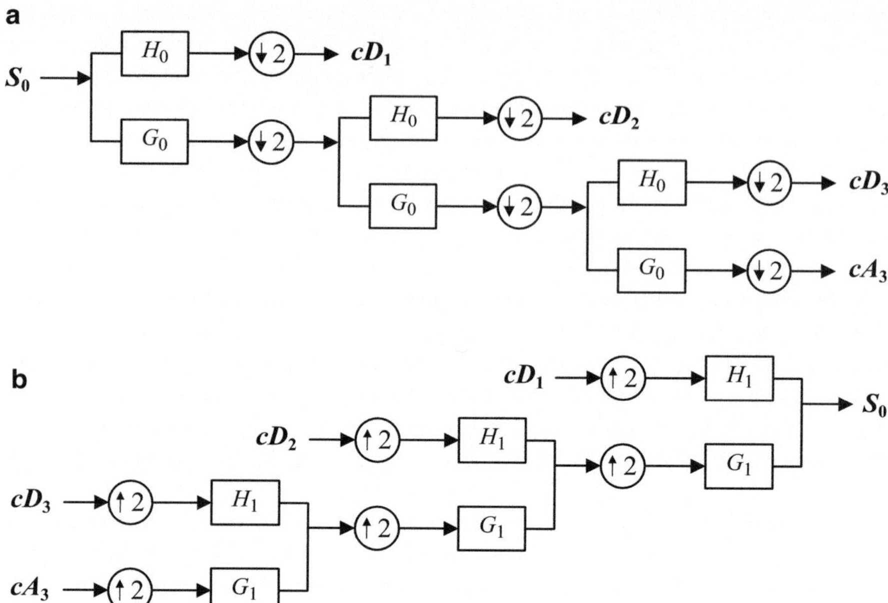

Fig. 3.9 A three-level DWT decomposition and reconstruction. (**a**) Wavelet decomposition. (**b**) Wavelet reconstruction.

Notes: 1. ⊕2 and ⊕2 denote downsampling and upsampling by two, respectively.

2. H_0/G_0 and H_1/G_1 are high-pass/low-pass analysis and synthesis filters, respectively

sounds, DWT exhibits similar characteristics in time–frequency resolution as the human ear [101]. As for the reconstruction, the original signal can be perfectly reconstructed from the DWT coefficients when analysis and synthesis filters satisfy the following conditions [102]:

$$H_0\,(-z)\,G_0\,(z) + H_1\,(-z)\,G_1\,(z) = 0,$$
$$H_0\,(z)\,G_0\,(z) + H_1\,(z)\,G_1\,(z) = 2. \tag{3.13}$$

Moreover, if both high- and low-frequency parts are iteratively decomposed, DWT is evolved into discrete wavelet packet transform (DWPT), which offers a more complex and flexible analysis of audio signals. For example, DWPT-based decomposition is employed for psychoacoustic modelling in [103].

3.2.5.1   Algorithm

Considering the good time–frequency resolution property of DWT, some audio watermarking schemes in the wavelet domain are proposed, such as [7, 101, 104–107]. Both methods in [101, 104] belong to wavelet domain watermarking using the SS method. In [104], a perceptually shaped PRS was taken as the watermark and spread over the detail coefficients at each level. To keep the imperceptibility and robustness, Hwang et al. [101] embedded the watermark into post-masking regions with high-energy and low zero-crossing rate (ZCR). Instead of DWT coefficients, certain detail coefficients of the last level from the DWPT were chosen for watermarking in [105]. Different from [7, 101, 104–106] embedded the watermark in wavelet domain using the patchwork method. Moreover, to enhance the security, Cvejic et al. [7] employed a secret key to randomly select the subbands used for watermarking.

Inspired by the idea of cepstrum domain watermarking in [95], Li et al. [107] applied SMM in the wavelet domain, where the mean of approximation coefficients at the last level is modified to embed the watermark. Basically, the procedures of watermark embedding and detection described in [107] are similar to Algorithm 3.2. To further improve the performance, we also employ the Hanning window and half overlapping for smooth transition as well as variable frame length to strive against PITSM. However, the estimation of detection threshold T_d is not necessary in wavelet domain watermarking based on SMM. Similar pre-attack experiments as those in cepstrum domain watermarking show that the detection threshold T_d can be specified as $T_d = 0$.

MATLAB script for wavelet domain watermarking can be found as $Wavelet-$ $watermarking.m$ file under Audio_Watermarking_Techniques folder in the attached CD.

3.2.5.2   Performance Evaluation

Figure 3.10 shows an example of the watermarked signal by wavelet domain watermarking based on SMM. In our implementation, a three-level DWT decomposition with "db4" wavelet [107] is carried out. Accordingly, the mean of DWT coefficients cA_3 is modified to embed the watermark. Considering the trade-off between imperceptibility and robustness, the watermarking strength α_w is set to be 0.01 or 0.02.

Comparing Table 3.5 with Table 3.4, wavelet domain watermarking based on SMM performs slightly better than cepstrum domain watermarking. However, DA/AD conversion is still a disaster to the survival of watermarks. As indicated by high ODG scores, the watermarked audio signals are perceptually undistinguished from the host signal, and informal subjective listening tests have also proved this. From Table 3.5, one point to note is that high SNRs are not equivalent to high ODG scores. The watermarked signals under the condition of $N = 1,024$ and $\alpha_w = 0.02$ have lower SNRs, but higher ODG scores than those of $N = 2,048$ and $\alpha_w = 0.01$. Seeing that the SNR is not a reliable indicator of perceptual quality for audio watermarking, more objective quality measures are investigated in Chap. 6.

Similarly, repetition coding can greatly improve the detection of wavelet domain watermarking based on SMM. When $\alpha_w = 0.01$ and $N = 1,024$, the watermarks generated with triple repetition coding are already able to combat various attacks from noise addition to Compression I at different bitrates, excluding DA/AD conversion. The resulting BERs are less than 5 %. Moreover, by using $N = 2,048$ and variable frame length (indicated as $\tilde{N}$), the watermarked signals are quite robust

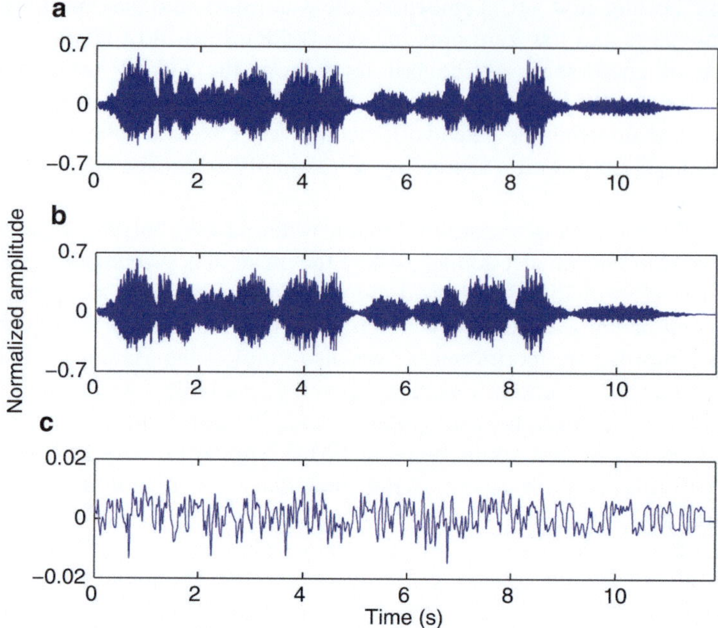

Fig. 3.10 Host signal and a watermarked signal by wavelet domain watermarking. Note that the watermarked signal is produced by watermarking with $N = 2,048$, $n_r = 3$, and $\alpha_w = 0.01$. (**a**) Host audio signal. (**b**) Watermarked audio signal. (**c**) Difference between the watermarked and host audio signals

against all the attacks except DA/AD conversion, cropping, and inserting. To further reduce the BERs under cropping and inserting, a longer frame (e.g., $N = 4,096$) would be required.

As a kind of statistical watermarking, wavelet domain watermarking based on SMM is also troubled with security. In [95], the watermark was encrypted before embedding for the purpose of security consideration. In this way, the encrypted watermark remains incomprehensible to the attacker without the secret key for decryption. However, the encryption only offers additional security on top of watermarking, but cannot prevent deliberate alteration on the mean of DWT coefficients from seriously destroying the embedded watermark. Therefore, it is quite important to enhance the security of the watermarking scheme.

3.2.6 Echo Hiding

Echo hiding embeds the watermark into host signals by introducing different echoes. With well-designed amplitudes and delays (offset), the echoes are perceived as resonance to host audio signals and would not produce uncomfortable noises [36].

Table 3.5 Results of performance evaluation of wavelet domain watermarking

	Watermarking parameters								
Frame length N	1024						2048		
Embedding strength α_w	0.01			0.02			0.01		
Watermark length N_w	1,016	338	203	1,016	338	203	504	168	100
Repetition coding n_r	1	3	5	1	3	5	1	3	5
(1) Perceptual quality assessment									
SNR/dB	23.38	22.92	22.90	20.80	20.01	20.06	26.28	25.90	25.59
ODG	0.15	0.15	0.16	0.10	0.13	0.12	−0.48	−0.47	−0.50
(2) Robustness test (BER: %)									
No attack	22.34	2.07	0	16.34	0.30	0	18.45	0	0
Noise (3 dB)	22.74	2.37	0	16.34	0.59	0	19.05	0	0
Resampling (22.05 kHz)	22.83	2.37	0	16.93	0.30	0	18.25	0	0
Requantization (8 bit)	22.44	2.07	0	16.14	0.30	0	19.25	0	0
Amplitude +10 %	22.34	2.07	0	16.34	0.30	0	18.45	0	0
−10 %	22.34	2.07	0	16.34	0.30	0	18.45	0	0
Lp filtering (8 kHz)	21.75	2.07	0	16.14	0.30	0	18.65	0	0
DA/AD (line-in jack)	47.24	47.04	50.74	47.83	47.93	46.80	48.41	48.21	55.91
Echo (0.3, 200 ms)	25.79	4.44	3.45	19.69	1.18	0	21.63	1.79	0
Reverb (1 s)	23.33	4.44	0.99	15.26	0.30	0	19.25	1.79	0

(continued)

Table 3.5 (continued)

		Watermarking parameters								
Frame length N		1024						2048		
Embedding strength α_w		0.01			0.02			0.01		
Compression I	96 kbps	22.05	2.66	0.49	17.03	0.30	0	19.05	0.60	0
	64 kbps	22.34	2.66	0	17.03	0.30	0	18.65	0.60	0
	48 kbps	21.95	2.96	0.49	16.83	0.59	0	18.85	0	0
Cropping (4×25 ms)		48.03	38.76	35.96	44.19	39.94	33.50	37.70	32.14	19.00
Jittering (0.1/20 ms)		46.85	41.72	33.00	44.49	31.36	21.67	38.69	22.02	2.00
Inserting (4×25 ms)		46.36	42.90	36.45	42.62	41.42	33.99	40.67	35.71	18.00
PITSM +10 %	N	48.62	47.04	44.33	49.70	48.82	46.31	50.40	54.76	46.00
	$\tilde{N}$	37.30	24.26	12.81	37.11	13.31	6.40	28.77	4.17	0
−10 %	N	49.52	49.03	46.77	51.45	54.52	42.47	52.04	47.10	45.16
	$\tilde{N}$	38.29	21.30	7.88	34.25	9.17	2.96	30.95	6.55	1.00
TPPSM +10 %		37.40	18.05	7.88	35.04	12.43	2.96	25.99	8.33	2.00
−10 %		32.87	14.50	6.90	29.63	4.44	1.97	26.79	4.17	1.00

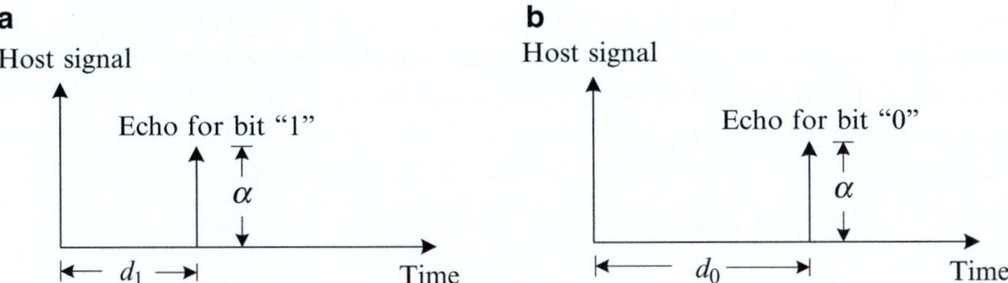

Fig. 3.11 Impulse response of echo kernels. (**a**) "One" kernel. (**b**) "Zero" kernel

3.2.6.1   Algorithm

In the embedding process, the watermarked signal $s_w(n)$ is generated by the convolution between host signal $s_o(n)$ and echo kernel $h(n)$. The basic echo hiding scheme employs a single echo kernel, whose impulse response is expressed as

$$h(n) = \delta(n) + \alpha\delta(n - d), \tag{3.14}$$

where α is echo amplitude and d is the delay. To represent bit "1" and "0," echo kernels are created with different delays (d_1 and d_0), as shown in Fig. 3.11. Usually, the allowable delay offsets for 44.1 kHz sampled audio signals are set to be $100 \sim 150$ samples (about $2.3 \sim 3.4$ ms) [108]. Consequently, the watermarked signal is described as follows:

$$\begin{aligned} s_w(n) &= s_o(n) \otimes h(n) \\ &= s_o(n) + \alpha \cdot s_o(n - d). \end{aligned} \tag{3.15}$$

In order to detect the watermark, cepstrum analysis is utilized to discern the value of delay. According to Fig. 3.6a, the complex cepstrum of the watermarked signal $\hat{s}_w(n)$ is defined as

$$\hat{s}_w(n) = \mathrm{F}^{-1}\left\{\log\left(\mathrm{F}\left\{s_w(n)\right\}\right)\right\}, \tag{3.16}$$

where $\mathrm{F}\{\cdot\}$ and $\mathrm{F}^{-1}\{\cdot\}$ denote the Fourier transform and the inverse Fourier transform, respectively. After substituting Eq. (3.15) into Eq. (3.16), $\hat{s}_w(n)$ is written as

$$\begin{aligned} \hat{s}_w(n) &= \mathrm{F}^{-1}\left\{\log\left(\mathrm{F}\left\{s_o(n) \otimes h(n)\right\}\right)\right\} \\ &= \mathrm{F}^{-1}\left\{\log\left(\mathrm{F}\left\{s_o(n)\right\}\right)\right\} + \mathrm{F}^{-1}\left\{\log\left(\mathrm{F}\left\{h(n)\right\}\right)\right\} \\ &= \mathrm{F}^{-1}\left\{\log\left(S_o\left(e^{j\omega}\right)\right)\right\} + \mathrm{F}^{-1}\left\{\log\left(H\left(e^{j\omega}\right)\right)\right\} \\ &= \hat{s}_o(n) + \hat{h}(n), \end{aligned} \tag{3.17}$$

where $\hat{s}_o(n) = \mathrm{F}^{-1}\{\log(S_o(e^{j\omega}))\}$ and $\hat{h}(n) = \mathrm{F}^{-1}\{\log(H(e^{j\omega}))\}$ are the complex cepstrum of $s_o(n)$ and $h(n)$, respectively. In view of Eq. (3.14), we have $H(e^{j\omega}) = 1 + \alpha e^{-j\omega d}$. Using the Taylor series $\log(1+x) = x - \frac{x^2}{2} + \frac{x^3}{3} - \cdots$ for $|x| < 1$, $\hat{h}(n)$ is calculated by

$$\begin{aligned}
\hat{h}(n) &= \mathrm{F}^{-1}\left\{\log\left(1 + \alpha e^{-j\omega d}\right)\right\} \\
&= \mathrm{F}^{-1}\left\{\alpha e^{-j\omega d} - \frac{\alpha^2}{2}e^{-j2\omega d} + \frac{\alpha^3}{3}e^{-j3\omega d} - \cdots\right\} \\
&= \alpha\delta(n-d) - \frac{\alpha^2}{2}\delta(n-2d) + \frac{\alpha^3}{3}\delta(n-3d) - \cdots
\end{aligned}$$
$$\tag{3.18}$$

Accordingly, $\hat{s}_w(n)$ in Eq. (3.17) becomes

$$\hat{s}_w(n) = \hat{s}_o(n) + \alpha\delta(n-d) - \frac{\alpha^2}{2}\delta(n-2d) + \frac{\alpha^3}{3}\delta(n-3d) - \cdots \tag{3.19}$$

This shows that a series of impulses with exponentially decaying amplitudes repeatedly appear for every d samples. In particular, the dominant spike is just located at the delay ($n = d$) and its amplitude is equal to that of the embedded echo, α. Then, the watermark can be decided based on the comparison between the values of cepstrum coefficients at two delays, i.e., $\hat{s}_w(d_1)$ and $\hat{s}_w(d_0)$.

To further increase the amplitude of cepstrum spikes representing the echoes, autocorrelation of the cepstrum (auto-cepstrum) $c_a(n)$ is employed to detect the delay [36, 109]:

$$c_a(n) = \mathrm{F}^{-1}\left\{\log(\mathrm{F}\{s_w(n)\})^2\right\}. \tag{3.20}$$

Since the autocorrelation calculates the signal power at each delay, the power spike in the cepstrum is more prominent, as illustrated in Fig. 3.12. Therefore, the watermark bit is determined by comparing $c_a(d_1)$ and $c_a(d_0)$:

$$w_e(i) = \begin{cases} 1, & \text{if } c_a(d_1) \geq c_a(d_0), \\ 0, & \text{if } c_a(d_1) < c_a(d_0). \end{cases} \tag{3.21}$$

The performance of echo hiding depends on echo kernels, and hence different echo kernels are introduced to improve the imperceptibility and robustness of the embedded echoes [3, 108, 110–112]. In [108], the echo kernel comprises multiple positive and negative echoes with different delays. Typically, a dual echo kernel with one positive and one negative echo is denoted as

$$h(n) = \delta(n) + \alpha_1\delta(n-d) - \alpha_2\delta(n-d-\Delta), \tag{3.22}$$

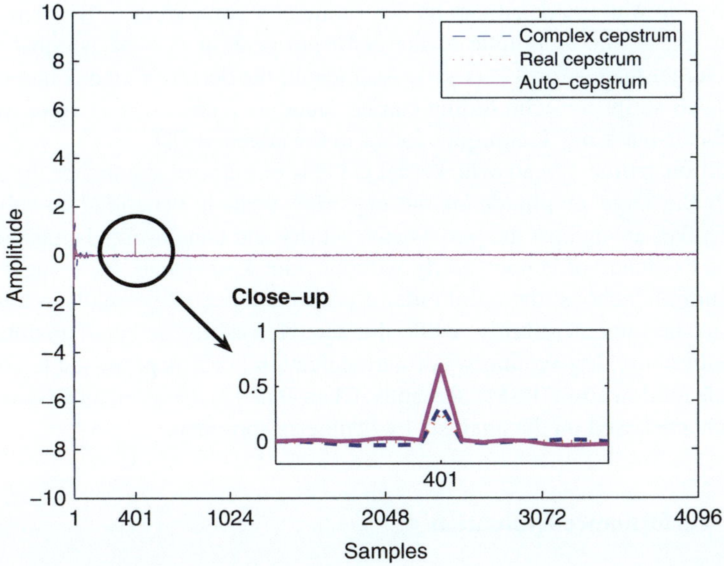

Fig. 3.12 Auto-cepstrum for echo detection

where $|\triangle| \leq 5$ samples. Then, its cepstrum $\hat{h}(n)$ is calculated as

$$\hat{h}(n) = \alpha_1 \delta(n - d) - \alpha_2 \delta(n - d - \triangle) + \frac{\alpha_1^2 + \alpha_2^2}{2} \{\delta(n - 2d) + \delta(n - 2d - 2\triangle)\}$$
$$+ \alpha_1 \alpha_2 \delta(n - 2d - \triangle) + \cdots \tag{3.23}$$

Therefore, the watermark bit can be determined by comparing $[c_a(d_1) - c_a(d_1 + \triangle)]$ and $[c_a(d_0) - c_a(d_0 + \triangle)]$. By virtue of the positive and negative echo kernel, high-energy echoes can be added to enhance the robustness, while audio quality is not deteriorated. This is because by combining these closely located positive and negative echoes, the frequency response of the dual echo kernel can remain flat over lower frequencies. Thus, the perceptual quality of the watermarked signal is preserved.

Later, Kim et al. [110] proposed the backward and forward echo kernel as follows:

$$h(n) = \delta(n) + \alpha \delta(n - d) + \alpha \delta(n + d). \tag{3.24}$$

Then, its cepstrum $\hat{h}(n)$ is calculated as

$$\hat{h}(n) = \alpha \{\delta(n - d) + \delta(n + d)\} - \frac{\alpha^2}{2} \{\delta(n - 2d) + 2\delta(n) + \delta(n + 2d)\}$$
$$+ \frac{\alpha^3}{3} \{\delta(n - 3d) + 3\delta(n - d) + 3\delta(n + d) + \delta(n + 3d)\} - \cdots$$
$$= \left(\alpha + \alpha^3 + \alpha^5 + \cdots\right)\delta(n - d) + \cdots$$
$$= \frac{\alpha}{1 - \alpha^2}\delta(n - d) + \cdots. \tag{3.25}$$

Therefore, the watermark bit can be determined by comparing $c_a (d_1)$ and $c_a (d_0)$. From Eq. (3.25), the amplitude of the cepstrum peak at $n = d$ is equal to $\frac{\alpha}{1-\alpha^2}$, which is larger than α for $0 < \alpha < 1$. As a result, the detection rate is increased.

MATLAB script for echo hiding can be found as $Echo - hiding.m$ file under Audio_Water-marking_Techniques folder in the attached CD.

In addition, a time-spread echo kernel [111] is introduced to enhance the security. Although the large amplitude of the cepstrum peak is beneficial to robustness, obvious spikes are against the purpose of security and unauthorized attackers might detect the existence of echoes easily without prior knowledge. By using a PRS to spread multiple echoes, the amplitude of each echo becomes small. It contributes directly to the imperceptibility, while the detection ability is better maintained as well. Furthermore, log-scaling watermark detection [112] is proposed to cope with pitch-scale modification (PSM). Recently, Chen et al. [3] designed an advanced echo hiding scheme based on the analysis-by-synthesis approach.

3.2.6.2 Performance Evaluation

In our experiments, echo hiding schemes with a single echo kernel (Kernel 1), a positive and negative echo kernel (Kernel 2), and a backward and forward echo kernel (Kernel 3) are evaluated. For a fair comparison, the frame length and the amplitude of different echo kernels are the same, i.e., $N = 4,096$ and $\alpha = 0.2$. In order to enhance the security, a sequence of pseudorandom numbers is utilized as the secret key to shift between several echo delays. Each delay is denoted as d_{xy}, where x and y represents the pseudorandom number (PRN) and the watermark bit, respectively. In this way, if the PRN is 1 and bit "0" is to be embedded, then d_{10} is selected. In considering both imperceptibility and robustness, the value of delays is set as follows: $d_{11} = 100$ and $d_{01} = 120$ are used for embedding bit "1," while $d_{10} = 110$ and $d_{00} = 130$ are used for embedding bit "0." Moreover, additional delay $\Delta = 4$ is used in the positive and negative echo kernel. Similar to previous watermarking techniques, Hanning windowing and half overlapping for smooth transition, variable frame length to combat PITSM and repetition coding are also employed to further improve the performance.

Figure 3.13 shows the watermarked signal produced by echo hiding using the positive and negative echo kernel. The results of performance evaluation of three echo kernels are summarized in Table 3.6. The positive and negative echo kernel provides higher SNRs than the other two kernels, which means the least distortion between the watermarked and host signals. However, the watermarked signals with three echo kernels obtain the same ODG scores and are deemed to be similar in perceptual quality. Informal subjective listening tests show that the added echoes do not introduce annoying noises, rather they make the sound rich.

Regarding the robustness of the watermarked signals, the backward and forward echo kernel (Kernel 3) generally provides the best detection rate. With the help of triple repetition coding ($n_r = 3$) and variable frame length (indicated as $\tilde{N}$), the BERs of watermarks under all attacks except TPPSM are less than 10 %. After

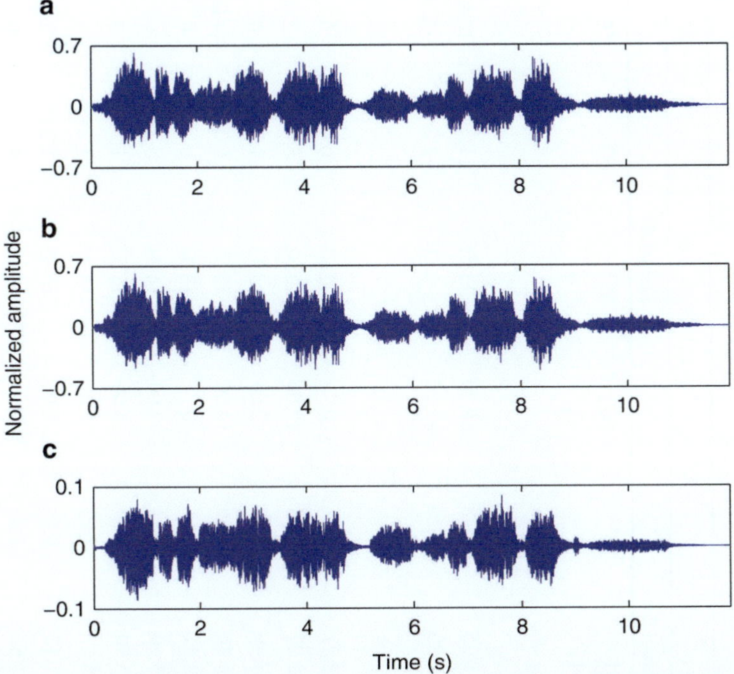

Fig. 3.13 Host signal and a watermarked signal by echo hiding. Note that the watermarked signal is produced by watermarking with $N = 4,096$, $\alpha = 0.2$, and $n_r = 3$. (**a**) Host audio signal. (**b**) Watermarked audio signal. (**c**) Difference between the watermarked and host audio signals

quintuple repetition coding ($n_r = 5$), the BERs under cropping and inserting are reduced further. For the other two kernels, the positive and negative echo kernel (Kernel 2) performs slightly better than the single echo kernel (Kernel 1), where the BERs are decreased on average. In addition, it is worth mentioning that all three kernels exhibit good resistance to DA/AD conversion. However, all the watermark detections with any kernel fail completely under TPPSM. Also, echo addition might be a hazardous attack on echo hiding watermarking. If echo delays in the attack happen to be the same as those of echo kernels, the mistakes in watermark detection are unavoidable.

Echo hiding is a watermarking technique specifically for audio signals. By selecting the proper amplitude and delay of echo kernels, the echoes embedded as the watermark can be imperceptible and robust against most attacks. However, echo hiding may suffer from two deficiencies. One is weak security, because obvious cepstrum peaks might be tampered with deliberately. The other is about inborn echoes contained in natural sound, which might result in false-positive errors [110].

Table 3.6 Results of performance evaluation of echo hiding

Echo hiding method	Watermarking parameters								
	Kernel 1			Kernel 2			Kernel 3		
Watermark length N_w	251	83	50	251	83	50	251	83	50
Repetition coding n_r	1	3	5	1	3	5	1	3	5
(1) Perceptual quality assessment									
SNR /dB	14.78	14.19	14.13	19.25	18.48	18.30	11.56	11.45	10.87
ODG	−2.20	−2.20	−2.20	−2.20	−2.20	−2.20	−2.20	−2.20	−2.20
(2) Robustness test (BER: %)									
No attack	27.09	4.82	2.00	22.71	2.41	0	12.75	0	0
Noise (30 dB)	35.86	21.69	14.00	32.27	20.48	12.00	23.51	6.02	2.00
Resampling (22.05 kHz)	24.70	9.64	2.00	27.49	6.02	2.00	17.13	1.20	0
Requantization (8 bit)	33.47	18.07	16.00	35.46	14.46	14.00	23.90	6.02	4.00
Amplitude +10 %	27.09	3.61	2.00	22.31	3.61	0	13.94	0	0
−10 %	27.89	3.61	2.00	23.90	2.41	2.00	14.34	0	0
Lp filtering (8 kHz)	28.29	9.64	2.00	25.90	8.43	4.00	15.54	1.20	0
DA/AD (line-in jack)	23.11	12.05	4.00	20.72	4.82	2.00	12.35	2.41	0
Echo (0.3, 200 ms)	28.69	7.23	8.00	21.12	8.43	4.00	16.33	1.20	2.00
Reverb (1 s)	24.70	6.02	2.00	21.51	4.82	0	15.14	0	0

Compression I	96 kbps		26.69	6.02	6.00	24.30	3.61	0	17.53	1.20	0
	64 kbps		23.51	10.84	2.00	20.32	3.61	0	17.13	1.20	0
	48 kbps		29.88	9.64	2.00	23.90	8.43	0	17.13	2.41	0
Cropping (4 × 25 ms)			41.83	19.28	6.00	38.25	14.46	6.00	32.67	9.64	2.00
Jittering (0.1/20 ms)			37.45	16.87	10.00	31.87	10.84	4.00	25.10	3.61	0
Inserting (4 × 25 ms)			39.04	15.66	4.00	35.46	18.07	4.00	28.69	8.43	2.00
PITSM	+4 %	N	47.81	43.37	28.00	47.81	34.94	24.00	42.23	37.35	36.00
		$\tilde{N}$	18.73	6.02	2.00	22.71	6.02	0.00	16.33	2.41	0
	−4 %	N	45.69	38.96	21.74	44.83	36.36	23.91	47.84	45.45	23.91
		$\tilde{N}$	23.90	8.43	0.00	24.30	2.41	2.00	15.54	4.82	2.00
TPPSM	+4 %		44.22	38.55	40.00	43.82	50.60	34.00	45.02	32.53	36.00
	−4 %		46.22	42.17	48.00	59.36	80.72	80.00	46.22	43.37	50.00

3.2.7 Histogram-Based Watermarking

As opposed to the previous methods, histogram-based watermarking works at
the global rather than the local characteristic of host signals, i.e., modifying the
histogram to embed the watermark. Histograms are well suited to describe the
distribution of large data sets. In digital watermarking, the histogram is obtained
by dividing a range of sample values into equal-sized bins and then calculating the
number of samples occurring in each bin. The samples used might be from the entire
or only a part of the host signal.

3.2.7.1 Algorithm

There are several ways in which the histogram is modified to embed the watermark.
Coltuc et al. [113] implemented a robust image watermarking using exact histogram
specification, where the image histogram is shape-altered (e.g., a saw-teeth shape
with $3 \sim 8$ periods) to represent a watermark. Later, MeÅŸe et al. [114] designed
an optimal algorithm for histogram modification, so that the mean square error
(MSE) is minimized between the modified and host images. In [115], the watermark
was embedded into the image by permuting some pairs of histogram bins. Apart
from being robust against geometrical attacks, the watermarking scheme is also
reversible, which means that the watermarked image can be fully restored to its
original status. Unlike the commonly used histogram in the time domain, Xuan
et al. [116] employed histogram shifting in the integer wavelet transform domain
for reversible image watermarking.

In [35], histogram modification is applied in audio watermarking. Based upon
the fact that the modified audio mean and the audio histogram shape are invariant
to temporal scaling, the authors designed an audio watermarking scheme resistant
to time-scale modification (TSM), random cropping, and inserting attacks. The
modified audio mean $\overline{A}$ is defined as the average of the absolute value of the
16-bit signed audio signal, i.e., $\overline{A} = \frac{1}{N_o} \sum_{n=1}^{N_o} |s_o(n)|$. In the embedding, $\overline{A}$ is used
to decide the amplitude range B of the samples for producing the histogram, i.e.,
$B = \left[-\lambda\overline{A}, \lambda\overline{A}\right]$. Through extensive experiments on different audio signals, a
suggested range is $\lambda \in [2, 2.5]$[35]. The number of histogram bins N_{bin} depends
on watermark length N_w, i.e., $N_{bin} \geq 3N_w$. The factor 3 comes from the reason
that the watermark is embedded by controlling the relative relation of the number
of samples in every three neighboring bins. Accordingly, the bin width N_{bw} is
calculated as $N_{bw} = \left\lfloor \frac{2\lambda\overline{A}}{N_{bin}} \right\rfloor$. The value of N_{bw} affects the properties of both
imperceptibility and robustness. A small N_{bw} is likely to reserve the shape of the
original histogram, which is beneficial to the perceptual quality. Meanwhile, each
bin should contain sufficient samples to ensure the watermark robustness. Therefore,
λ must be carefully chosen to obtain a suitable N_{bw}.

After the histogram is constructed, its bins are divided into N_w groups, each of which has three bins. For every group, the number of samples in three consecutive bins is denoted as N_{b1}, N_{b2}, and N_{b3}, respectively. Then, one watermark bit is embedded into one group of histogram bins by applying the following rules [117]:

$$
\begin{cases}
\dfrac{(N_{b1}+N_{b3})}{2N_{b2}} \geq E_h, & \text{if } w_o\,(i) = 1, \\[2mm]
\dfrac{2N_{b2}}{(N_{b1}+N_{b3})} \geq E_h, & \text{if } w_o\,(i) = 0,
\end{cases}
\tag{3.26}
$$

where the embedding strength E_h is around $1.2 \sim 1.5$. Obviously, a large E_h would increase the robustness but degrade the perceptual quality.

From the embedding process, one point to note is that histogram-based watermarking has the advantage of handling silent intervals of the host audio signal. As mentioned in Sect. 3.1.1, it is nearly impossible to embed the watermark into zero values. In histogram-based watermarking, all zero-value samples fall into the center of the histogram. Therefore, these samples can be well preserved, provided that the one or two bins right in the center are exempted from watermarking.

In the detection, the modified mean of the watermarked signal $\overline{A}^{w}$ is calculated, and then the histogram is constructed with the same λ and N_{bin}. Furthermore, the histogram bins are divided into groups in the same way as the embedding. Therefore, the watermark bit is decided by comparing the number of samples in three consecutive bins, namely, N_{b1}^{w}, N_{b2}^{w}, and N_{b3}^{w}:

$$
w_e\,(i) =
\begin{cases}
1, & \text{if } \dfrac{(N_{b1}^{w}+N_{b3}^{w})}{2N_{b2}^{w}} \geq 1, \\[3mm]
0, & \text{if } \dfrac{(N_{b1}^{w}+N_{b3}^{w})}{2N_{b2}^{w}} < 1.
\end{cases}
\tag{3.27}
$$

Since the attacks might change the modified audio mean, $\overline{A}^{w}$ is probably not equal to $\overline{A}$. As a result, the histogram range is deviated and the watermark cannot be detected correctly. Therefore, it is necessary to search for the proper modified audio mean. Experimental results show that the fluctuation of the modified audio mean is usually less than $\pm 6\,\%$; thus, optimal searching within this range is proposed to prevent exhaustive searching [35].

MATLAB script for histogram-based watermarking can be found as $Histogram-watermarking.m$ file under Audio_Watermarking_Techniques folder in the attached CD.

3.2.7.2 Performance Evaluation

Figure 3.14 shows an example of the watermarked signal by histogram-based watermarking[117]. The difference between the watermarked and host audio signals is small, and accordingly the SNRs in Table 3.7 are fairly high. From Fig. 3.14c, one point to note is that most samples in the front and rear part of the host signal are changed to embed the watermark, but the middle part remains intact. This is

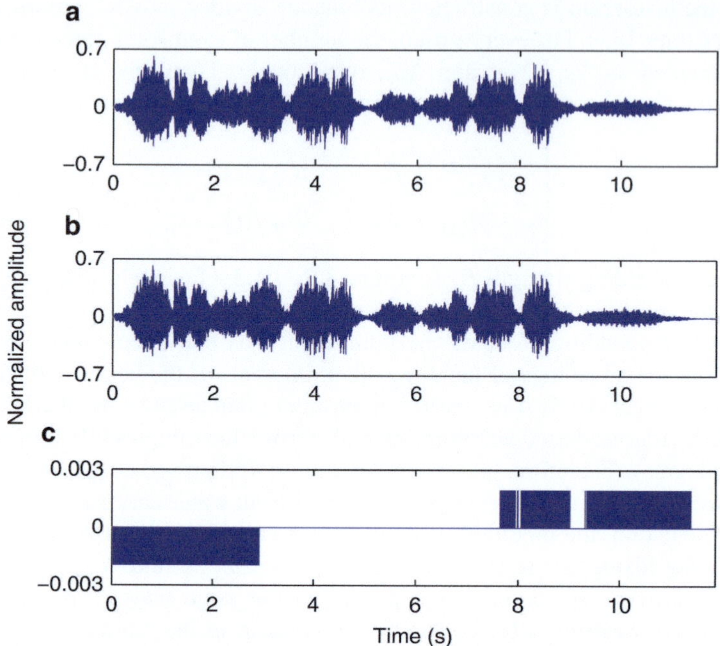

Fig. 3.14 Host signal and a watermarked signal by histogram-based watermarking. Note that the watermarked signal is produced by watermarking with $N_w = 40$, $\lambda = 2.2$, and $E_h = 1.5$. (**a**) Host audio signal. (**b**) Watermarked audio signal. (**c**) Difference between the watermarked and host audio signals

because the samples in each histogram bin are arranged in the order they are found in the host signal. During the process of modifying the samples into three bins of each group, the samples at the beginning and the end of each bin are always chosen with priority. Our implementation attempted to randomly select the samples in the bins, but the perceptual quality of the watermarked signal became worse. The reason might be that the randomly distributed modification resembles the addition of white noise.

The results of performance evaluation of histogram-based watermarking are illustrated in Table 3.7. Generally, a larger N_w or a smaller λ would result in a better perceptual quality, but a weaker robustness. This is because the bin width N_{bw} is proportional to λ, but inversely proportional to N_w. Moreover, histogram-based watermarking is indeed quite robust against most desynchronization attacks such as TSM, cropping, jittering, and inserting. In compromise with the imperceptibility, the watermarks are also able to survive from PSM to a certain extent.

However, the BERs of the watermarks are rather high in the cases of some attacks like low-pass filtering, DA/AD conversion, reverberation, and MP3 compression. Even extending the searching range of $\overline{A}^w$ to $\pm 10\%$, the detection rate has no substantial improvement.

Table 3.7 Results of performance evaluation of histogram-based watermarking

	Watermarking parameters							
Watermark length N_w	20				40			
Embedding range λ	2.2		2.5		2.2		2.5	
Embedding strength E_h	1.2	1.5	1.2	1.5	1.2	1.5	1.2	1.5
(1) Perceptual quality assessment								
SNR /dB	41.99	38.66	40.96	37.55	47.95	44.59	46.70	43.35
ODG	−2.17	−2.89	−2.21	−2.86	−1.73	−2.42	−1.83	−2.60
(2) Robustness test (BER: %)								
No attack	0	0	0	0	0	0	0	0
Noise (30 dB)	5.00	0	0	0	20.00	12.50	25.00	17.50
Resampling (22.05 kHz)	0	0	0	0	0	0	0	0
Requantization (8 bit)	30.00	30.00	35.00	35.00	35.00	35.00	27.50	25.00
Amplitude +10 %	0	0	0	0	0	0	0	0
−10 %	0	0	0	0	0	0	0	0
Lp filtering (8 kHz)	5.00	5.00	5.00	0	27.50	22.50	22.50	17.50
DA/AD (line-in jack)	25.00	20.00	35.00	30.00	32.50	30.00	27.50	30.00
Echo (0.3, 200 ms)	25.00	30.00	30.00	30.00	15.00	22.50	20.00	22.50
Reverb (1 s)	25.00	25.00	15.00	15.00	27.50	22.50	17.50	20.00

(continued)

Table 3.7 (continued)

		Watermarking parameters							
Watermark length N_w		20				40			
Embedding range λ		2.2		2.5		2.2		2.5	
Compression I	9 kbps	0	0	0	0	15.00	15.00	10.00	15.00
	64 kbps	5.00	5.00	5.00	5.00	15.00	17.50	22.50	20.00
	48 kbps	15.00	5.00	15.00	10.00	22.50	25.00	22.50	27.50
Cropping (4 × 25 ms)		0	0	0	0	0	0	0	0
Jittering (0.1/20 ms)		0	0	0	0	0	0	0	0
Inserting (4 × 25 ms)		0	0	0	0	0	0	0	0
PITSM	+10 %	0	0	0	0	0	0	0	0
	−10 %	0	0	0	0	0	0	0	0
TPPSM	+10 %	0	0	0	0	10.00	0	7.50	5.00
	−10 %	0	0	0	0	2.50	0	5.00	2.50

The reason is that the attacks have smoothed the histograms and the relative relations between N_{b1}^w, N_{b2}^w, and N_{b3}^w of each group are destroyed. As an exception, the histogram changes dramatically after requantization. Since quantization rounds off the samples, some histogram bins are eliminated, which leads to failure in watermark detection. For an 8-bit quantizer, the quantization error is calculated as $q = \frac{2\cdot\max(s_o(n))}{2^8} = \frac{\max(s_o(n))}{128}$ [118]. Therefore, the width of histogram bins should be larger than q to combat requantization attack.

Later, instead of the histogram in the time domain, Xiang et al. [2] exploited the invariance of histogram shape in the low-frequency subband of the DWT domain. It is reported that the improved watermarking method is more robust against TSM, low-pass filtering, and MP3 compression. Nevertheless, the security of the watermarking scheme is still a serious issue.

In addition to the mean, other statistical moments can be used as invariants in histogram-based watermarking. For example, image steganalysis in [119] employed the first four moments (i.e., the mean, variance, skewness, and kurtosis) of the subbands that are decomposed by quadrature mirror filters (QMF). In the book, we attempted to utilize the moments in [120] for audio watermarking, where the audio signal has been converted into a two-dimensional square matrix as implemented in [121]. However, most invariants (except for the first moments) are not applicable to audio watermarking. The reason might be that the audio signal represented in a two-dimensional form is incomparable to a two-dimensional image.

3.3 Summary

This chapter has mainly investigated different audio watermarking techniques, such as LSB modification, phase coding, SS watermarking, cepstrum domain watermarking, wavelet domain watermarking, echo hiding, and histogram-based watermarking. We also contributed to the improvement of some of these techniques.

To achieve a better comprehension of various methods, specifications on the execution of performance evaluation were described. Firstly, seventeen pieces of audio test signals in four categories were prepared for evaluation. The diversity of test signals helps verify the applicability of audio watermarking techniques. Secondly, the instructions for conducting perceptual quality assessment including subjective listening tests and objective evaluation tests were presented. For subjective listening tests, a group of trained participants are required to grade the watermarked signals in MUSHRA test and the SDG rating. For objective evaluation tests, both the ODG provided by software PEAQ and the SNR are adopted as impersonal measurements. Thirdly, test items and their default parameters in basic and advanced robustness tests were separately depicted in detail. A number of common signal operations, desynchronization attacks, and advanced attacks are included in the robustness test.

Following that, different audio watermarking techniques were intensively studied. LSB modification and phase coding are early audio watermarking techniques.

Although LSB modification might be unperceived, both methods performed relatively poorly in the robustness test. SS watermarking is the most prevalent technique for digital watermarking because of its robustness and security. However, effective ways for SS watermarking to combat desynchronization attacks are required. The implemented cepstrum domain watermarking and wavelet domain watermarking were based on SMM. After our improvement, their performance in terms of imperceptibility and robustness became fairly good. However, the issue of security is a challenge for such statistical watermarking. Echo hiding is a method especially for audio watermarking. With the proper echo kernel, the echoes embedded as the watermark can be made imperceptible and resistant to most attacks. Different from the previous techniques, histogram-based watermarking is rather robust against desynchronization attacks, but somewhat vulnerable to several common signal processing operations, such as low-pass filtering and reverberation.

Two observations can be obtained from the experiments. One is that trade-offs always exist between imperceptibility, robustness, security, and data payload in audio watermarking systems. The other is that there are no universal audio watermarking techniques to combat all the attacks. In most cases, it is more difficult to conquer desynchronization attacks.

Chapter 4
Proposed Audio Watermarking Scheme

Imperceptibility, robustness, and security are vital considerations in the design of any audio watermarking scheme for copyrights protection. In this chapter, a spread spectrum (SS)-based audio watermarking technique which involves the psychoacoustic model, multiple scrambling, adaptive synchronization, frequency alignment, and coded-image watermark is presented. To preserve the perceptual quality of the watermarked signal, amplitude shaping using the psychoacoustic model is employed. Also, the proposed scheme integrates multiple scrambling operations into the embedding process to prevent unauthorized detection. That is, the amount and position of the slots used for embedding each watermark bit are randomly set and certain subbands are randomly selected for the embedding. Moreover, adaptive synchronization and frequency alignment are developed to retrieve the watermarks from the attacked watermarked signals that suffer loss of synchronization. In addition, the information to be embedded can be encrypted with a coded-image, so as to provide a semantic meaning for verification as well as extra security.

The chapter is organized as follows. Section 4.1 describes the selection of watermarking regions and the structure of the watermarking domain. In Sect. 4.2, the embedding algorithm including multiple scrambling is presented. Section 4.3 focuses on watermark detection emerging with adaptive synchronization and frequency alignment. This is followed by an introduction to the coded-image watermark in Sect. 4.4. Finally, Sect. 4.5 summarizes the characteristics of the proposed audio watermarking scheme.

4.1 Preliminaries

Generally, the proposed audio watermarking scheme includes watermark embedding and watermark detection. In watermark embedding, we embed not only the watermark for copyrights protection, but also synchronization information for watermark detection. These data are basically represented by the bits. In our scheme,

Y. Lin and W.H. Abdulla, *Audio Watermark: A Comprehensive Foundation Using MATLAB*,   95
DOI 10.1007/978-3-319-07974-5_4, © Springer International Publishing Switzerland 2015

the watermark bits are repeatedly embedded to increase the robustness. Also, a synchronization bit is repeatedly embedded for synchronization purposes.

For a better understanding of the watermarking procedure, we start with some preliminary knowledge, such as the selection of watermarking regions and the structure of the watermarking domain.

4.1.1 Selection of Watermarking Regions

As mentioned in Sect. 3.1.1, embedding the watermarks into silent segments would introduce unavoidable perceived noise. Therefore, a selection process is applied on the host audio signal to determine the embedding regions. Correspondingly, it is necessary to perform this procedure in the watermark detection stage, so as to locate the regions for watermark detection. Since various attacks might alter the watermarked signal, the watermarking regions should be rather stable to ensure that they still can be identified in the attacked signals [4, 5, 122]. In this way, the process of selection is a kind of initial synchronization between the watermark embedding and detection.

There are several methods for selecting reliable watermarking regions which usually follow certain distinct points, for example, salient point extraction [6, 33, 97], peak point extraction [123], and envelope peak extraction [5]. Commonly, these delicate methods have been employed as solutions to synchronization; however, this is not necessary in the proposed scheme. The selection of our watermarking regions mainly aims to preclude the long silences from watermarking, not to solve the synchronization problem for watermark detection. Thus, the accuracy of locating watermarking regions is not required to be as high as that in synchronization.

Similar to the method in [4], the proposed scheme selects the watermarking regions according to the signal energy. The energy is calculated on a frame-by-frame basis along the input signal, where each non-overlapping frame $g(n)$ has a length of N. Then, successive frames whose energy, i.e., $\sum_{n=1}^{N} g^2(n)$, exceed a certain threshold E_T are concatenated to construct a high-energy segment. In our scheme, a high-energy segment which has a duration of more than 2 s is considered to be a long high-energy segment. Furthermore, adjacent long high-energy segments which are located within 0.1 s are concatenated into one watermarking region.

The predefined energy threshold E_T plays an important role in the selection of watermarking regions. For different values of E_T, different watermarking regions are selected from the input signal, as shown in Fig. 4.1.

With a lower E_T, more audio frames are included and hence longer segments will be available for embedding the watermark. In this case, data payload is higher. However, since the segments with relative low energy are susceptible to attacks, the obtained watermarking regions might be unstable. With a higher E_T, more stable segments will be chosen for watermarking, but data payload is reduced accordingly. Therefore, given the watermark to be embedded, E_T is then determined to achieve a better compromise between the robustness and data payload.

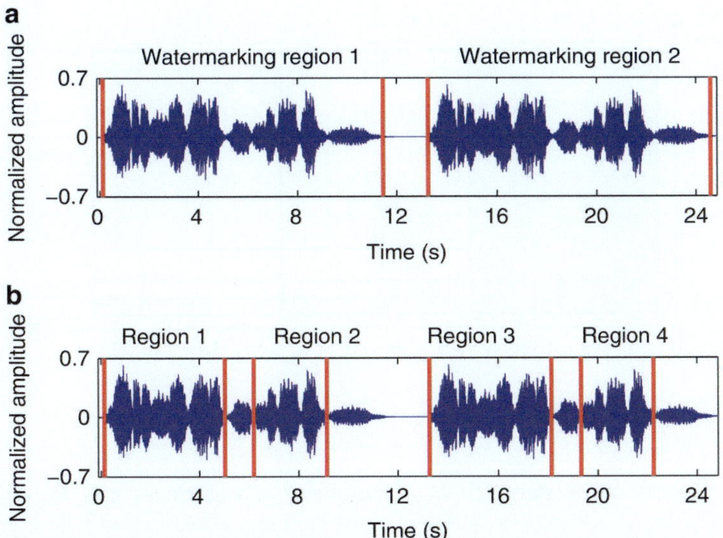

Fig. 4.1 Selection of watermarking regions (**a**) $E_T = 0.01$ (**b**) $E_T = 0.1$

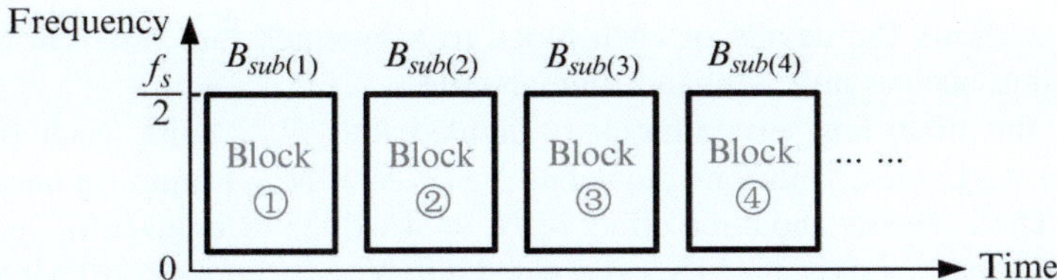

Fig. 4.2 Diagram of blocks in the watermarking domain

4.1.2 Structure of the Watermarking Domain

The watermarking domain, which is generated by taking the fast Fourier transform (FFT) of adjacent audio frames with 50 % overlap, refers to the time–frequency representation of the selected watermarking regions. Each frame has a length of N points. As shown in Fig. 4.2, the watermarking domain is divided into N_{block} blocks. The reason for the half overlapping is to smooth the transition between frames, as used by the previous techniques in Chap. 3.

Each block is used to embed one sub-watermark B_{sub}, which is a part of the original watermark w_o, i.e.,

$$w_o = \left\{ B_{\text{sub}(m)} \right\} \quad m = 1, \ldots, N_{\text{block}}. \tag{4.1}$$

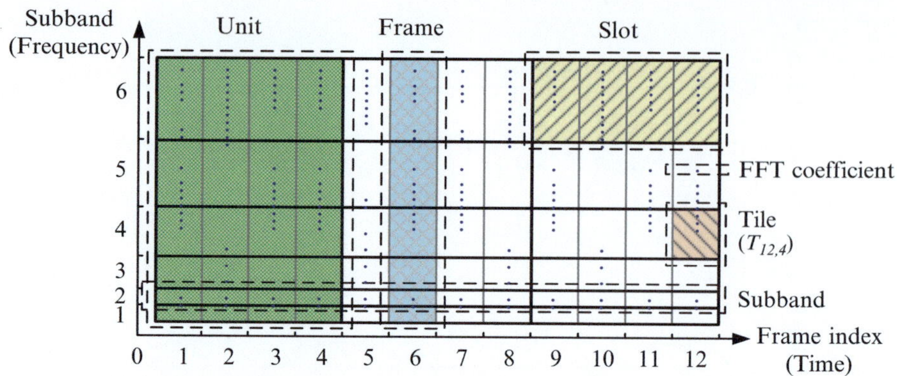

Fig. 4.3 Configuration of one block

Moreover, every sub-watermark B_{sub} contains N_{bit} watermark bits $B_i \in \{1, -1\}$, i.e.,

$$B_{\text{sub}} = \{B_i\} \quad i = 1, \ldots, N_{\text{bit}}. \tag{4.2}$$

Figure 4.3 shows the details of each block as segmented into different levels of granularities, such as unit, subband, slot, and tile.

Along the time axis, every block is divided into N_{unit} units, each of which comprises N_c frames. Thus, one block has $N_f = N_c \times N_{\text{unit}}$ frames. In our scheme, $N_c = 4$. The concerns about the effect of $N_c = 4$ will be recognized in Sect. 4.3.1. Along the frequency axis, the block is divided into N_{subband} nonuniform perceptually motivated subbands based on the Gammatone filterbank (GTF) (to be described in Sect. 4.1.3).

The intersection of a subband and a unit is called a slot. Each watermark bit $B_i \in \{1, -1\}$ is repeatedly embedded into a number of slots for robustness purposes. In addition to the watermark bits, a synchronization bit $B_s = 1$ is also repeatedly embedded for synchronization purposes. Specifically, N_B slots are randomly chosen from the total number of slots within every block for embedding each B_i. Then the remaining slots are used for embedding B_s, whose total number is N_s. The value of N_s is calculated by

$$N_s = N_{\text{unit}} \times N_{\text{subband}} - N_{\text{bit}} \times N_B. \tag{4.3}$$

In this way, every slot is used for embedding a bit, as exemplified in Fig. 4.4. Without loss of generality, it is assumed here that each sub-watermark consists of two watermark bits, i.e., $B_{\text{sub}} = \{B_1, B_2\}$, which are separately embedded into five slots, i.e., $N_B = 5$. Thereby, $N_s = 3 \times 6 - 2 \times 5 = 8$ slots are used for embedding B_s.

To enhance watermarking security, every bit embedded in the slot is modulated by a pseudorandom number (PRN), $P_x \in \{1, -1\}$. As mentioned above, N_B slots

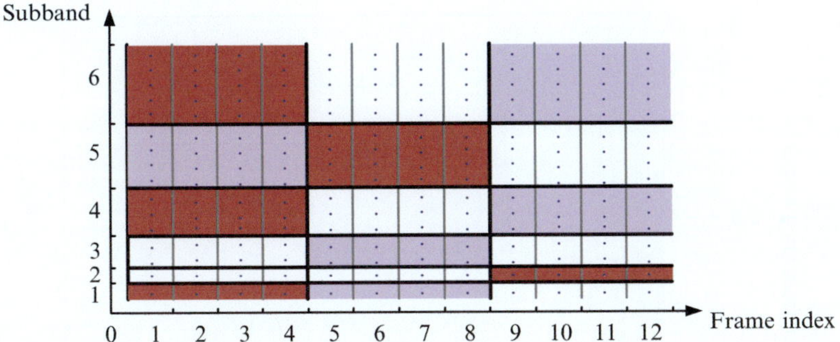

Fig. 4.4 Distribution of the watermark bits and synchronization bit. *Note*: Slots, ▇, ▢ and ▢ are used for embedding B_1, B_2, and B_s, respectively

are chosen for embedding each B_i. Accordingly, these N_B slots correspond to N_B PRNs, denoted by $P_{b_i} = \{P_{b_i}(k)\}$, $k = 1, 2, \ldots, N_B$. Similarly, N_s slots for embedding B_s correspond to a total of N_s PRNs, denoted by $P_s = \{P_s(k)\}$, $k = 1, 2, \ldots, N_s$.

The distribution of the bits in all blocks is determined by one secret key k_b, which belongs to confidential information shared only between the embedder and the authorized detectors. According to k_b, different bits are spread on every block, as exemplified in Fig. 4.4. To clearly describe the distribution of the bits, we build up a bit matrix M_B. Based on the structure of the block, M_B is a N_{subband} by N_{unit} matrix. A PRN matrix M_P of the PRNs is constructed correspondingly. The PRN matrix M_P is of the same size as M_B, i.e., $N_{\text{subband}} \times N_{\text{unit}}$. For example, M_B and M_P for the block shown in Fig. 4.4 are expressed as

$$
M_B = \begin{bmatrix} B_1 & B_s & B_2 \\ B_2 & B_1 & B_s \\ B_1 & B_s & B_2 \\ B_s & B_2 & B_s \\ B_s & B_s & B_1 \\ B_1 & B_2 & B_s \end{bmatrix} \quad \text{and} \quad M_P = \begin{bmatrix} P_{b_1}(3) & P_s(5) & P_{b_2}(5) \\ P_{b_2}(1) & P_{b_1}(4) & P_s(8) \\ P_{b_1}(2) & P_s(4) & P_{b_2}(4) \\ P_s(2) & P_{b_2}(3) & P_s(7) \\ P_s(1) & P_s(3) & P_{b_1}(5) \\ P_{b_1}(1) & P_{b_2}(2) & P_s(6) \end{bmatrix} . \tag{4.4}
$$

In this example, M_P consists of $P_{b_1} = \{P_{b_1}(1), P_{b_1}(2), \ldots, P_{b_1}(5)\}$, $P_{b_2} = \{P_{b_2}(1), \ldots, P_{b_2}(5)\}$, and $P_s = \{P_s(1), P_s(2), \ldots, P_s(8)\}$, which correspond to B_1 (in red), B_2 (in blue), and B_s (in black), respectively. Without loss of generality, our implemented watermark detection (to be described in Sect. 4.3) searches B_s and each B_i which are embedded in the block, from the left bottom and then column by column. Correspondingly, the indices of P_s and P_{b_i} in Eq. (4.4) separately start from the left bottom and column by column.

Note that all blocks have the same configuration of M_B, which is solely determined by k_b. But the value of M_B varies from block to block, since each block has different watermark bits $\{B_i\}$, $i = 1, \ldots, N_{\text{bit}}$ to be embedded. As for M_P,

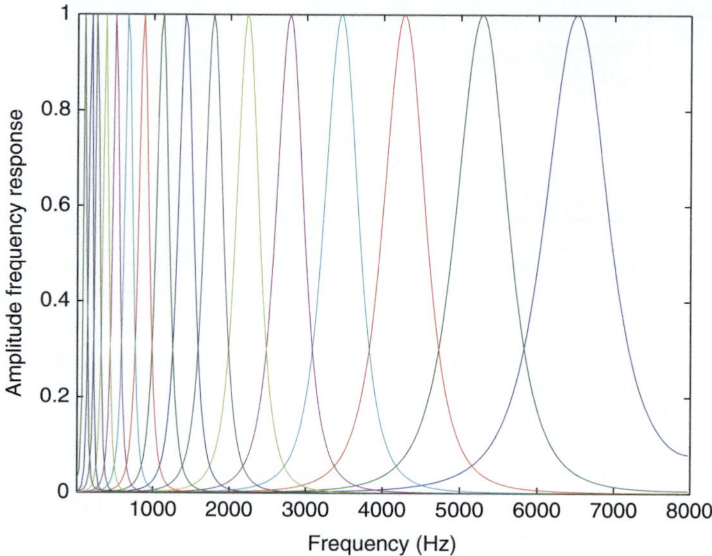

Fig. 4.5 Frequency response of a 16-channel GTF

it is unique and the same as all blocks to keep synchronization. The values of M_P components are determined by one secret key k_p, which also belongs to confidential information shared only between the embedder and the authorized detectors.

The smallest basic element of the slot is called tile as indicated in Fig. 4.3. A tile consisting of several FFT coefficients is the basic module for amplitude modulation in the watermark embedding. Due to $N_c = 4$, each slot contains four tiles. Recall that every slot is used for embedding one bit, which is modulated by a PRN. More specifically, the four tiles in a slot are used for embedding a bit and they share the PRN of that slot in common. Accordingly, the tiles used for embedding each bit are identified in the watermark detection and used to determine the bit value.

Each tile is denoted as $T_{t,b}$, where t and b are its frame and subband indices respectively. For example, $T_{12,4}$ shown in Fig. 4.3 stands for the tile located at the 4th subband of the 12th frame.

4.1.3 Gammatone Auditory Filterbank

As mentioned in Sect. 4.1.2, the Gammatone filterbank (GTF) is employed for setting the subbands. The GTF is a bank of overlapping band-pass filters, which mimics the frequency response of the human cochlea, wider bandwidths at higher frequencies and narrower bandwidths at lower frequencies [124]. For the sake of illustration, Fig. 4.5 shows an example of the frequency response of a 16-channel GTF, covering 100–8,000 Hz frequency band.

To get the center frequency and the bandwidth of each subband, the entire carrying band ($f_L \sim f_H$) and the number of channels N_{GTF} are required. Then, overlapping spacing υ is calculated by

$$\upsilon = \frac{9.26}{N_{\text{GTF}}} \log \left(\frac{f_H + 228.7}{f_L + 228.7} \right), \tag{4.5}$$

where f_L and f_H are lower and upper frequency limits in Hz.

So the nth ($1 \leq n \leq N_{\text{GTF}}$) subband's center frequency (f_c) and bandwidth (B_w) in Hz are computed by

$$f_c = -228.7 + (f_H + 228.7) \cdot \exp \left(-\frac{\upsilon n}{9.26} \right) \tag{4.6}$$

$$B_w = 24.7 \left(1 + 4.37 f_c \right). \tag{4.7}$$

However, the channels from the GTF are usually overlapped, which results in confusion of embedding bits. In order to get a set of non-overlapping subbands, the lower limit of each channel (V^l) is always taken as the boundary, where $V^l = f_c - B_w/2$. This is because critical bandwidths are determined by the lower edges of the band [11]. Moreover, narrow channels in the low frequencies might contain only one FFT coefficient or even less under a sampling frequency of 44.1 kHz. Since a single frequency coefficient is sensitive towards slight modification, several channels in the low frequencies are combined in our scheme. In this way, the tiles are forced to contain more than five FFT coefficients, which greatly helps improve the robustness. For example, a set of 32 nonuniform subbands (i.e., $N_{\text{subband}} = 32$) over the frequency spectrum is illustrated in Appendix G.

4.2 Watermark Embedding

Watermark embedding is to modulate the amplitude of the host audio signal in the watermarking domain by a certain information used for copyrights protection [125]. In this section, the embedding algorithm integrated with multiple scrambling is described in detail.

4.2.1 Embedding Algorithm

For a given host signal to be watermarked, the first step of the embedding algorithm is the selection of the watermarking regions. This is followed by the construction of the watermarking domain. Generally, the watermark bits are embedded through amplitude modulation of the tiles in the watermarking domain.

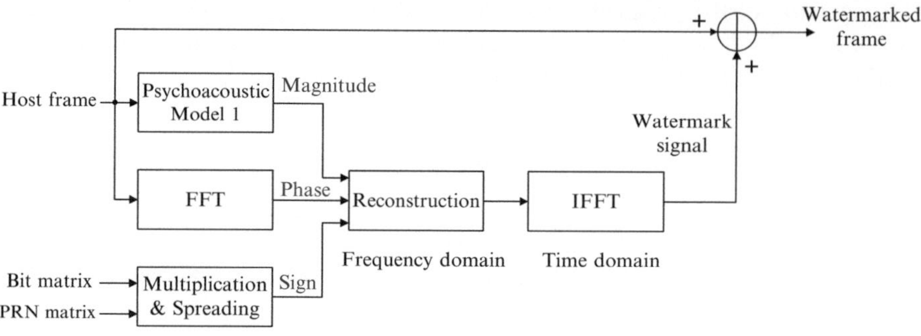

Fig. 4.6 Block diagram of watermarking one host frame

For each host frame, an imperceptible watermark signal is constructed in the frequency domain using the magnitude, phase, and sign of the signal spectrum. Then it is added to the host frame after being inversely transformed to the time domain. Figure 4.6 shows the block diagram of watermarking one host frame.

The magnitude spectrum of the watermark signal is determined by Psychoacoustic Model 1. As discussed in Sect. 2.4, the input to the psychoacoustic model is one host frame and the output is the minimum masking threshold (MMT). Therefore, the noise introduced by the watermarking is kept inaudible to human ears.

It is worth mentioning that the MMT shown in Fig. 2.24 cannot be directly used as the magnitude spectrum, because sound pressure level (SPL) normalization is involved in the first step of implementing Psychoacoustic Model 1. Thus, we need to multiply the MMT by a scale factor called watermark strength α_w, in order to obtain a proper magnitude spectrum. The effect of α_w will be discussed in Sect. 5.1.

The phase of the watermark signal is the same as that of the host signal to avoid phase distortion.

The sign of the spectrum of the watermark signal depends on the watermark bits (i.e., bit matrix, M_B) and the corresponding PRNs (i.e., PRN matrix, M_P). As described in Sect. 4.1.2, every slot is used for embedding a bit, which is modulated by a PRN. So the sign of a slot is defined as the multiplication of its bit and PRN. Then, the sign of a slot is spread over its four tiles. Specifically, if the sign of a slot is positive, the sign of the first two tiles in that slot is positive and that of the last two tiles is negative, and vice versa. Furthermore, the FFT coefficients in a tile share the sign of that tile. In this way, all the FFT coefficients in the blocks obtain their own signs.

After the calculation of the magnitude, phase, and sign, the frequency spectrum of the watermark signal is constructed. Recall from Sect. 3.2.2.1 that a real-valued signal has a conjugate symmetric spectrum. Thus, the construction process is performed as follows:

$\% F_{wm} (1 : N)$ is the frequency spectrum of the watermark signal.

$$\% \, F_p \left(1 : \frac{N}{2} \right) \text{ is the positive-frequency part of } F_{wm}.$$

$$\% \, F_n \left(1 : \frac{N}{2} \right) \text{ is the negative-frequency part of } F_{wm}.$$

$$F_p \left(1 : \frac{N}{2} \right) = \text{sign } .* \text{ magnitude } .* \exp(j*\text{phase});$$

$$F_n \left(1 : \frac{N}{2} \right) = \text{fliplr} \left(\text{conj} \left(F_p \left(1 : \frac{N}{2} \right) \right) \right);$$

$$F_{wm} (1 : N) = \left[F_p \left(1 : \frac{N}{2} \right), \, 0, \, F_n \left(1 : \frac{N}{2} - 1 \right) \right];$$

where N is the frame length, function fliplr $(\cdot)$ is to flip a matrix left to right, and function conj $(\cdot)$ is to compute the complex conjugate.

After taking the inverse fast Fourier transform (IFFT), the frequency spectrum of the watermark signal is transformed to the time domain and subsequently added to the host frame to produce the watermarked frame.

Finally, all the watermarked frames in each watermarking region are windowed by a Hanning window for smooth concatenation. By combining original samples with the watermarked regions in order, the overall watermarked audio signal is formed.

Figure 4.7 shows an example of the watermarked signal, where the host signal is the audio test file A_2 (*Bass.wav*). It is observed that the watermark signal has a similar shape to the host signal, which might help preserve the perceptual quality of the watermarked signal.

4.2.2 Multiple Scrambling

To increase the level of security, multiple scrambling can be employed in the embedding.

As described in Sect. 4.1.2, the amount and position of the slots used for embedding each watermark bit are randomly set, where two secret keys, i.e., k_b and k_p, are used. Furthermore, instead of using all subbands, we randomly select $\tilde{N}_{\text{subband}}$ out of N_{subband} subbands and randomize their orders of encoding, where two secret keys are used. So the number of possible embedding ways is calculated by the following permutation:

$$N_{\text{scrambling}} = P \left(N_{\text{subband}}, \tilde{N}_{\text{subband}} \right) = \frac{N_{\text{subband}}!}{\left(N_{\text{subband}} - \tilde{N}_{\text{subband}} \right)!}. \tag{4.8}$$

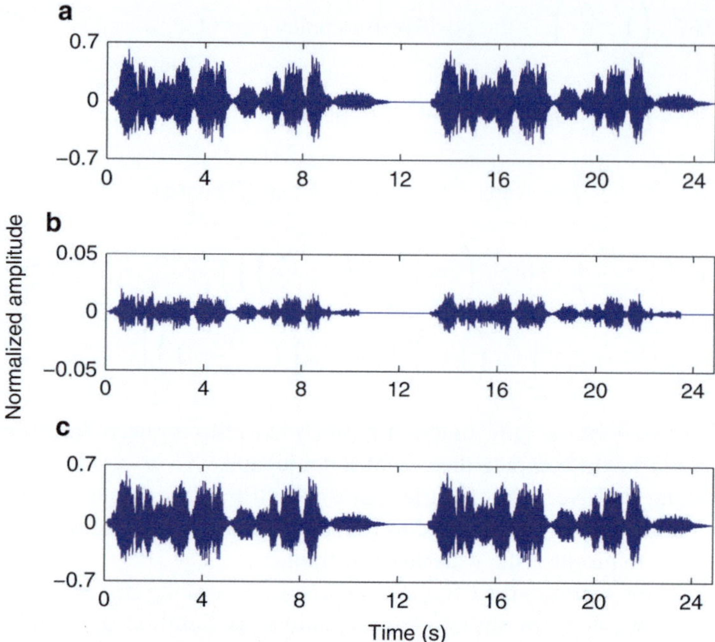

Fig. 4.7 Host signal and a watermarked signal by the proposed scheme. (**a**) Host signal. (**b**) Watermark signal. (**c**) Watermarked signal

Another possible scrambling operation is to encrypt the coded-image watermark (to be described in Sect. 4.4) into incomprehensible ciphers, where one secret key is used. Even if the watermark is extracted, the attacker still cannot recognize the image without the correct decipher.

Therefore, anyone without all the secret keys rarely has the ability to discern the embedded watermark. Since the secret keys are shared only between the embedder and authorized detectors, the aim of copyrights protection is really achieved. It is worth mentioning that public-key algorithms in modern communication systems [126] may be introduced to establish secure communication for sharing the secret keys.

4.3 Watermark Detection

From the description of the embedding algorithm in Sect. 4.2.1, the watermark bits in a bit matrix term are used to determine the sign of the spectrum of the watermark signal. Therefore, in the detection, every watermark bit is determined by checking whether it is an increase or decrease in the magnitudes of the corresponding tiles[1] of the attacked signal.[2] By correlating to the PRN matrix M_P defined in Sect. 4.1.2, we can detect the watermark without resorting to the host audio signal.

Since the watermark bits are embedded on a block-by-block basis, the beginning frame of each block is required by the watermark detection to identify the tiles used for embedding each bit. From the synchronization perspective, the beginning frame in each block denotes the synchronization position of that block. However, various attacks may modify the watermarked signals in different ways, and hence the synchronization positions are destroyed. Therefore, synchronization methods are required to find out the best synchronization position, which closely indicates the beginning of each block.

The synchronization in our scheme is initially achieved by block synchronization. As described in Sect. 4.1.2, a synchronization bit $B_s = 1$ is repeatedly embedded in each block. Therefore, the synchronization position is considered to be the location where the distribution of the tiles used for embedding B_s is matched.

However, as described in Sect. 4.1.2, the entire PRN matrix M_P and the configuration of the bit matrix M_B are the same for all blocks. Thus, in some cases when the match with the tiles used for embedding B_s is obtained, the synchronization position found by block synchronization is incorrect for detecting the watermark bits. By introducing a threshold T_{sync} to check whether a synchronization position can be accepted, adaptive synchronization [127] is developed to amend the block synchronization.

As concluded from the study on different audio watermarking techniques, it is more difficult to combat desynchronization attacks, especially pitch-invariant time-scale modification (PITSM) and tempo-preserved pitch-scale modification (TPPSM). Both can be implemented by audio editing tools without large perceptual impairment.[3] On most occasions, basic detection updated with adaptive synchronization is robust to a limited extent against PITSM and TPPSM. To efficiently handle excessive PITSM and TPPSM, frequency alignment [128] adjusts the frequency spectra that have been scaled by the attacks, so that the detection can retrieve the synchronization positions for recovering the embedded watermark.

Figure 4.8 shows the block diagram of watermark detection. The complete algorithm consists of three parts: basic detection, adaptive synchronization, and frequency alignment.

---

[1]As mentioned in Sect. 4.1.2, the tile is the basic module for amplitude modulation in the watermark embedding. Therefore, the detection is focused directly on the tiles, not the slots.

[2]As defined in Sect. 1.2.1, the input to the watermark detector is generally called the attacked signal, no matter whether it has been attacked or not. In the case that a watermarked signal has been attacked, we specifically call it an attacked watermarked signal.

[3]The random stretching attack used by [9, 10] which was implemented by omitting or inserting a random number of samples (usually called "random samples cropping/inserting") and the pitch shifting attack by linear interpolation are much less complicated than PITSM and TPPSM.

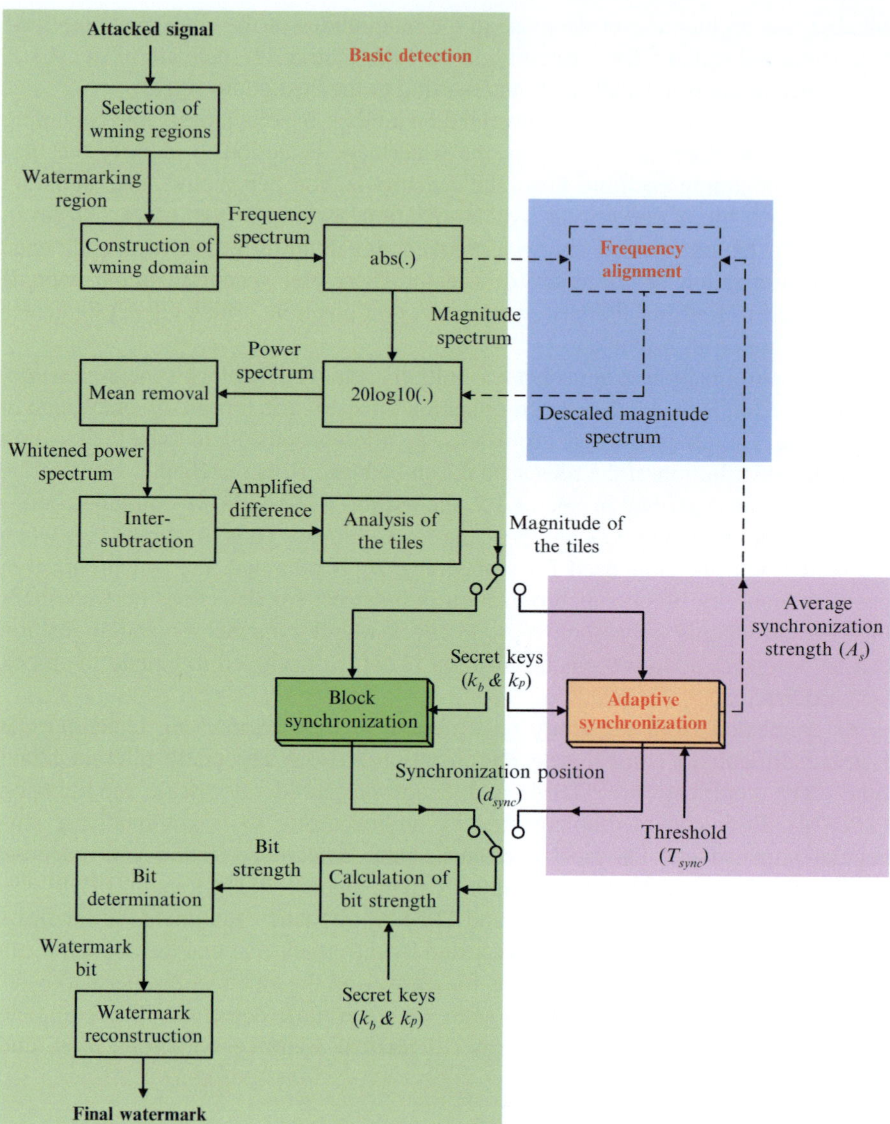

Fig. 4.8 Block diagram of watermark detection. *Notes*: (1) Basic detection works independently. (2) Adaptive synchronization is an improvement technique for block synchronization. (3) Frequency alignment indicated by dashed lines is an additional solution to excess PITSM and TPPSM

- Basic detection is the basic algorithm for watermark detection, which includes a set of steps for detecting the watermark from the attacked signal. It begins with the selection of watermarking[4] regions and the construction of the watermarking domain, similar to the embedding algorithm. Then, several operations are performed to calculate the magnitude of the tiles, which are required by block synchronization. By use of the secret keys, block synchronization aims to find out the synchronization position of each block. Based on the synchronization position found, the tiles used for embedding each watermark bit are identified and further utilized to calculate the bit strength. According to the bit strength, the value of that watermark bit is determined. Finally, the watermark bits detected from all the blocks comprise the whole watermark.

- Adaptive synchronization is an improvement technique for block synchronization. Except that a threshold T_{sync} is introduced as an extra input, adaptive synchronization has the same inputs (i.e., magnitude of the tiles and secret keys) and the same output (i.e., synchronization position d_{sync}) as block synchronization. Unless requested by frequency alignment, adaptive synchronization does not output the average synchronization strength (A_s).

- Frequency alignment is an additional solution to excessive PITSM and TPPSM of up to $\pm 10\%$. Only when basic detection updated with adaptive synchronization cannot detect the watermark from the attacked watermarked signal, frequency alignment is employed to descale the frequency spectra. Average synchronization strength from adaptive synchronization is involved in choosing the scale factor.

4.3.1 Basic Detection

As seen from Fig. 4.8, basic detection consists of the following eleven steps:

- **Step 1: Selection of the watermarking regions**

For a given signal, successive frames which exceed an energy threshold E_T are specified and concatenated into high-energy segments as watermark regions. Only these segments are used for watermark detection.

- **Step 2: Construction of the watermarking domain**

Similar to the approach in the embedding, the watermarking domain for detection is also generated by taking the FFT of adjacent frames with 50 % overlap. However, Hanning windowing is applied on these frames before taking the FFT. This is because all the watermarked frames are windowed and subsequently concatenated into the watermarked signal, as described in Sect. 4.2.1.

The frequency spectrum of the tth windowed frame, $F_t(n)$, is calculated by

$$F_t(n) = \text{FFT}\left\{g_t(n) \cdot \text{Hanning}(n)\right\} \quad 1 \leq n \leq N, \tag{4.9}$$

---

[4]The word "watermarking" is abbreviated as "wming" in the first two boxes in Fig. 4.8.

where N is the frame length, $g_t(n)$, $1 \leq n \leq N$ is the tth frame, and Hanning (n), $1 \leq n \leq N$ is the N-point Hanning window.

In view of the conjugate symmetry of the frequency spectrum, we only process the positive-frequency part with $\frac{N}{2}$ coefficients, i.e., $F_t\left(1:\frac{N}{2}\right)$.

- **Step 3: Calculation of the magnitude spectrum**

The magnitude spectrum of the tth windowed frame is shown to be $\left|F_t\left(1:\frac{N}{2}\right)\right|$, where $|\cdot|$ is the absolute value.

- **Step 4: Calculation of the power spectrum**

The power spectrum of the tth windowed frame, $P_t\left(1:\frac{N}{2}\right)$, is calculated by

$$P_t\left(1:\frac{N}{2}\right) = 20\log_{10}\left(\left|F_t\left(1:\frac{N}{2}\right)\right|\right). \qquad (4.10)$$

- **Step 5: Calculation of the whitened power spectrum**

The whitened power spectrum, $\hat{P}_t\left(1:\frac{N}{2}\right)$, is calculated by removing the mean from the power spectrum:

$$\hat{P}_t\left(1:\frac{N}{2}\right) = P_t\left(1:\frac{N}{2}\right) - \overline{P_t}, \qquad (4.11)$$

where $\overline{P_t} = \dfrac{1}{N/2}\sum_{n=1}^{N/2}[P_t(n)]$ is the mean of $P_t\left(1:\frac{N}{2}\right)$.

- **Step 6: Inter-subtraction**

To amplify the effect of the watermark signal and reduce the effect of the host signal, the difference $\tilde{P}_t\left(1:\frac{N}{2}\right)$ between each frame and the one just after the next is calculated:

$$\tilde{P}_t\left(1:\frac{N}{2}\right) = \hat{P}_t\left(1:\frac{N}{2}\right) - \hat{P}_{t+2}\left(1:\frac{N}{2}\right). \qquad (4.12)$$

The reason is that on the condition that adjacent frames in the blocks are half overlapped, these two frames (t and $t+2$) are considered to be non-overlapped with each other.

- **Step 7: Analysis of the tiles**

To accumulate the effect of the watermark signal over the FFT coefficients in a tile, the magnitude of each tile is calculated. Specifically, the magnitude of the tile located at the bth subband of the tth frame, $Q_{t,b}$, is calculated by

$$Q_{t,b} = \frac{\displaystyle\sum_{n=V_b^l}^{V_b^h} \tilde{P}_t(n)}{V_b^h - V_b^l + 1},\qquad(4.13)$$

where V_b^l and V_b^h refer to the lower and upper bounds of the bth subband, respectively.

- **Step 8: Block synchronization**

The purpose of block synchronization is to find out the beginning frame of each block.

Since every frame in its block is possibly the beginning frame, it is necessary to calculate synchronization strength S_d $(d = 1, \ldots, N_f)$ frame by frame, where $N_f = N_c \times N_{\text{unit}}$. On the assumption that the dth frame is the beginning frame, S_d for the dth frame is calculated by the following normalized correlation [13]:

$$S_d = \frac{\displaystyle\sum_{k=1}^{N_s}\left[Q_{t(d,k),b(k)}\cdot P_s(k)\right]}{\sqrt{\displaystyle\sum_{k=1}^{N_s}\left[Q_{t(d,k),b(k)}\right]^2}\cdot\sqrt{\displaystyle\sum_{k=1}^{N_s}[P_s(k)]^2}},\qquad(4.14)$$

where $\{Q_{t(d,k),b(k)}\}$ represent N_s tiles that are used for embedding B_s under the assumed condition and $\{P_s(k)\}$ are their corresponding PRNs. The subscripts $t(d,k)$ and $b(k)$ to locate the tiles are computed by

$$t(d,k) = d + [R_s(k,1) - 1] \times N_c \qquad(4.15)$$

$$b(k) = R_s(k,2), \qquad(4.16)$$

where N_c is the number of frames per unit and $N_c = 4$ is considered in our scheme. R_s is called the index matrix for B_s, which solely depends on the secret key k_p mentioned in Sect. 4.1.2. To indicate the location of these tiles, the first and second columns of R_s represent the distribution of the tiles' columns and rows, respectively. For example, given M_B in Eq. (4.4) which corresponds to Fig. 4.4, its R_s is shown as follows. As mentioned already in Eq. (4.4), the search for B_s starts from the left bottom and column by column without loss of generality.

$$
M_B = \begin{array}{c} 6 \\ 5 \\ 4 \\ 3 \\ 2 \\ 1 \end{array}
\begin{bmatrix}
B_1 & \underline{B_s} & B_2 \\
B_2 & B_1 & \underline{B_s} \\
B_1 & \underline{B_s} & B_2 \\
\underline{B_s} & B_2 & \underline{B_s} \\
\underline{B_s} & \underline{B_s} & B_1 \\
B_1 & B_2 & \underline{B_s}
\end{bmatrix}
\qquad \Longrightarrow \qquad
R_s = \begin{bmatrix}
1 & 2 \\
1 & 3 \\
2 & 2 \\
2 & 4 \\
2 & 6 \\
3 & 1 \\
3 & 3 \\
3 & 5
\end{bmatrix}
\tag{4.17}
$$

Then, the beginning frame of this block (i.e., the synchronization position d_{sync}) is the frame that provides the maximum S_d:

$$
d_{\mathrm{sync}} = \arg \max_{1 \le d \le N_f}(S_d). \tag{4.18}
$$

Also, the maximum S_d of that d_{sync} is denoted as $S_{d_{\mathrm{sync}}}$.

- **Step 9: Calculation of bit strength**

Based on the synchronization position found by Eq. (4.18), the bit strength of each watermark bit B_j, G_j, is calculated by

$$
G_j = \frac{\displaystyle\sum_{k=1}^{N_B}\left[Q_{t(d_{\mathrm{sync}},k),b(k)} \cdot P_{b_j}(k)\right]}{\sqrt{\displaystyle\sum_{k=1}^{N_B}\left[Q_{t(d_{\mathrm{sync}},k),b(k)}\right]^2} \cdot \sqrt{\displaystyle\sum_{k=1}^{N_B}\left[P_{b_j}(k)\right]^2}}, \tag{4.19}
$$

where the subscripts are computed by

$$
t(d_{\mathrm{sync}}, k) = d_{\mathrm{sync}} + \left[R_{b_j}(k,1) - 1\right] \times N_c \tag{4.20}
$$

$$
b(k) = R_{B_j}(k,2). \tag{4.21}
$$

Similarly, $\left\{Q_{t(d_{\mathrm{sync}},k),b(k)}\right\}$ and $\{P_{b_j}(k)\}$ refer to N_B tiles that are used for embedding B_j and their corresponding PRNs, respectively. As for R_{b_j}, it is the index matrix for B_j. Also, based on M_B of Fig. 4.4, R_{b_1} and R_{b_2} are built as follows:

$$
R_{b_1} = \begin{bmatrix}
1 & 1 \\
1 & 4 \\
1 & 6 \\
2 & 5 \\
3 & 2
\end{bmatrix}
\quad \text{and} \quad
R_{b_2} = \begin{bmatrix}
1 & 5 \\
2 & 1 \\
2 & 3 \\
3 & 4 \\
3 & 6
\end{bmatrix}.
$$

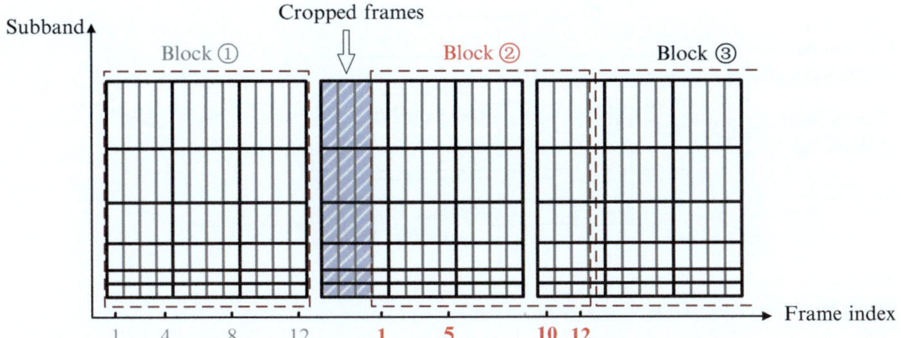

Fig. 4.9 Illustration of random samples cropping

- **Step 10: Determination of the watermark bit**

The value of watermark bit, B_j, is determined according to its bit strength G_j.

$$
\begin{aligned}
&\text{If } G_j \geq 0, \text{ then } B_j = 1. \\
&\text{If } G_j < 0, \text{ then } B_j = 0.
\end{aligned}
\tag{4.22}
$$

- **Step 11: Reconstruction of the watermark**

The watermark bits extracted from all the blocks are combined and the final watermark w_e is obtained.

4.3.2 Adaptive Synchronization

Adaptive synchronization is an improvement technique for block synchronization. As described in Sect. 4.3.1, the watermark bits are detected based on the synchronization position (d_{sync}) found by block synchronization. Recall from Sect. 4.1.2 that the distribution of the synchronization bit B_s (indexed by R_s) and the values of the corresponding PRNs (i.e., P_s) are the same for all blocks. Thus, in some cases, although the d_{sync} from Eq. (4.18) provides the best match with the tiles used for embedding B_s, it is a false position for detecting the watermark bits.

Take random samples cropping as an example of the attacks. Suppose that the first three frames in the second block have been cropped, as illustrated in Fig. 4.9.

Our task is to detect the watermark bits embedded in the second block. To this end, block synchronization is performed for this block, which means that the 12 frames indicated with red numbers are separately assumed to be the beginning frame. Due to the fact that these three blocks have the same R_s and P_s, the d_{sync} found from Eq. (4.18) would be the current tenth frame that is originally the first frame of the third block. Then, based on Eqs. (4.20) and (4.21), the tiles for bit

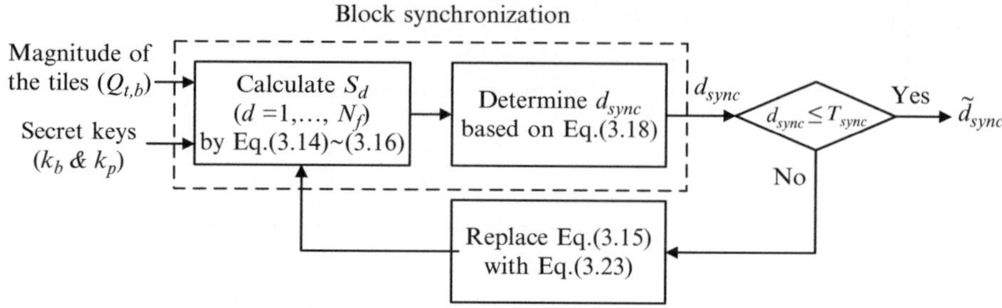

Fig. 4.10 Flowchart of adaptive synchronization

detection in Eq. (4.19) will be taken from the third block, instead of the second block. Consequently, the watermark bits embedded in the second block cannot be correctly detected.

To find out the actual synchronization position of the second block (i.e., the current first frame), we need to compensate for the cropped frames. According to the structure of the block, the number of the cropped frames can be calculated by $d_m = N_f - d_{\text{sync}} + 1 = N_c \times N_{\text{unit}} - d_{\text{sync}} + 1 = 4 \times 3 - 10 + 1 = 3$. Therefore, when identifying the tiles for embedding B_s, we take into consideration an offset of d_m.

Specifically, $t\,(d,k)$ in Eq. (4.15) should be modified into the following equation:

$$t\,(d,k) = d + [R_s\,(k,1) - 1] \times N_c - d_m, \qquad (4.23)$$

where $d_m = N_f - d_{\text{sync}} + 1$. Subsequently, new tiles are identified for block synchronization, in order to find out the actual synchronization position $\left(\tilde{d}_{\text{sync}}\right)$ of this block.

One point to note is that $t\,(d,k)$ calculated by Eq. (4.23) might be a negative value. For example, suppose that M_B in Eq. (4.4) is embedded in the second block in Fig. 4.9, then R_s is given in Eq. (4.17) and $d_m = 3$. According to Eq. (4.23), $t\,(1,1)$ and $t\,(1,2)$ are calculated by $t\,(1,1) = 1 + [R_s\,(1,1) - 1] \times 4 - 3 = -2$ and $t\,(1,2) = 1 + [R_s\,(2,1) - 1] \times 4 - 3 = -2$. In this case, these tiles cannot be identified and their values are simply replaced by zeros.

Furthermore, $t\,(d_{\text{sync}},k)$ in Eq. (4.20) for calculating G_j is also modified into

$$t\left(\tilde{d}_{\text{sync}},k\right) = \tilde{d}_{\text{sync}} + \left[R_{b_j}\,(k,1) - 1\right] \times N_c - d_m. \qquad (4.24)$$

The procedure described above is called adaptive synchronization. The key of adaptive synchronization is to choose a threshold $\left(T_{\text{sync}}\right)$ for determining whether a d_{sync} can be accepted. A d_{sync} is considered to be incorrect if it is larger than T_{sync}. Then, d_{sync} should be recalculated and eventually reach a value lower than T_{sync}, which will be taken as the $\tilde{d}_{\text{sync}}$. Figure 4.10 shows the flowchart of adaptive synchronization.

Note that if more than half of the frames in a block are missing or destroyed, we do not expect to detect the watermark bits embedded in that block. Therefore, $1 \leq d_m \leq \frac{N_f}{2}$ is considered. In view of $d_m = N_f - d_{\text{sync}} + 1$, then $\frac{N_f}{2} + 1 \leq d_{\text{sync}} \leq N_f$. This means that under different attacks, T_{sync} is varied between $\left(1 + N_f/2\right)$ and N_f. Thus, we need to perform adaptive synchronization for each possible T_{sync} and subsequently calculate the average synchronization strength $\left(A_{\text{sync}}\right)$ that is defined as

$$A_{\text{sync}} = \frac{1}{N_{\text{block}}} \sum_{k=1}^{N_{\text{block}}} \left[S_{\tilde{d}_{\text{sync}}}(k) \right], \qquad (4.25)$$

where $S_{\tilde{d}_{\text{sync}}}(k)$ is the kth block's $S_{\tilde{d}_{\text{sync}}}$. Then, the T_{sync} that provides the maximum A_{sync} is regarded as the desired one. Experimentally, an optimal value of T_{sync} is $\lfloor 0.8 N_f \rfloor$, where $\lfloor \cdot \rfloor$ is the smallest integer value.

4.3.3 *Frequency Alignment Towards Excessive PITSM and TPPSM*

Among various attacks, time-scale modification (TSM) and pitch-scale modification (PSM) are more likely to raise difficulties in the process of watermark detection. In most cases, adaptive synchronization can only cope with PITSM and TPPSM at a limited extent (within $\pm 4\,\%$). Although the requirement of the SDMI standard described in Appendix A has been satisfied, a higher level up to $\pm 10\,\%$ is desired by STEP 2000 described in Appendix B.

In order to combat an excessive distortion of PITSM and TPPSM, we suggest performing frequency alignment to adjust the frequency spectra that have been scaled. Hence, synchronization positions can be retrieved for recovering the embedded watermark from a severely attacked watermarked audio file.

4.3.3.1 Frequency Alignment Against TSM and PSM

Time- and pitch-scale modification of audio signals refer to the operations of independently controlling and modifying the time evolution and the pitch contour, respectively. In particular, PITSM is to slow down or speed up a given signal without altering its pitch. TPPSM, meanwhile, is to shift the pitch upward or downward without affecting the duration [129].

As their definitions imply, there is a duality between TSM and PSM [129]. A pitch-scaled signal can be obtained by a TSM followed by a sampling rate alteration. On the other hand, a time-scaled signal can be achieved by changing the sampling rate of a signal after PSM, as shown in Fig. 4.11. This shows that a solution to withstanding PSM is also applicable to TSM, since a time-scaled signal can be

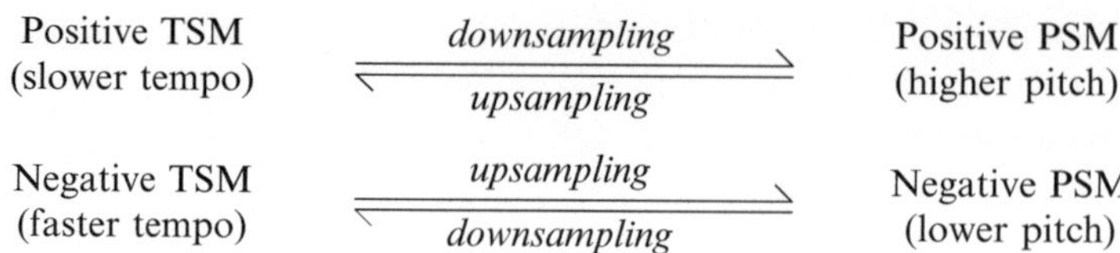

Fig. 4.11 Duality between TSM and PSM

converted to a pitch-scaled signal after an alteration of the sampling rate. Therefore, PSM is used for illustrating how to perform frequency alignment.

The operation of PSM can be formulated as

$$\tilde{f} = \alpha \cdot f, \tag{4.26}$$

where α (> 0) is the scale factor, f is the frequency of the original signal, and $\tilde{f}$ is the α-scaled frequency [112]. If $\alpha > 1$, it is a positive PSM that gets a higher pitch. If $\alpha < 1$, it is a negative PSM that gets a lower pitch. Accordingly, such a frequency fluctuation introduces desynchronization in watermark detection.

To retrieve the synchronization positions, frequency alignment attempts to reverse the process of PSM. Thus, the modified frequency spectrum is descaled by

$$f_{\text{alignment}} = \beta \cdot \tilde{f}, \tag{4.27}$$

where $\beta = 1/\alpha$. It appears as an operation of compression for positive PSM and expansion for negative PSM. In this manner, the original f could be recovered approximately.

4.3.3.2 Implementation of Frequency Alignment

The strategy for conducting frequency alignment in watermark detection is described as follows. After applying basic detection updated with adaptive synchronization on an attacked signal, we calculate the bit error rate (BER) between the original watermark (w_o) and the extracted watermark (w_e). If the BER is less than the threshold T_{BER}, the attacked signal under inspection is claimed to be a watermarked copy and the detection process is terminated [2]. If the BER is larger than T_{BER}, it is considered that the suspected signal might have been attacked by excessive TSM and/or PSM. Thus, frequency alignment is employed to descale the frequency spectra before the calculation of magnitude spectra, in an attempt to recover the synchronization positions.

It is worth mentioning that T_{BER} is not necessary to be an accurate value and merely provides a rough idea of whether the detection proceeds or not. Moreover, one valuable benefit offered by coded-image watermark (to be described in Sect. 4.4) is a semantic meaning of copyright information. The detection is

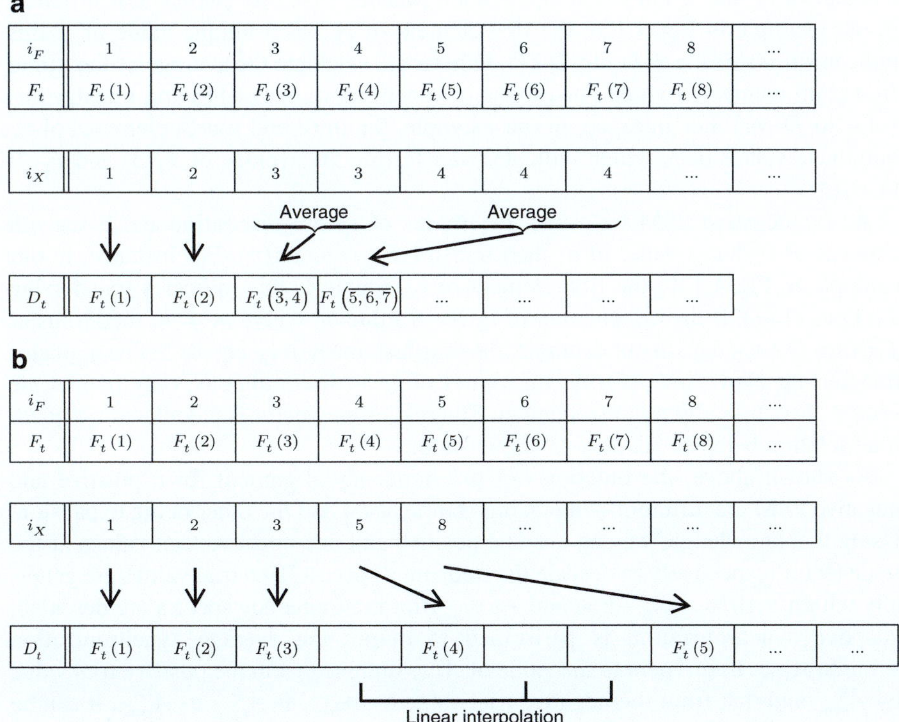

Fig. 4.12 Illustration of frequency alignment (**a**) Positive PSM (**b**) Negative PSM

terminated provided that the copyright information can be recognized. Even without knowing T_{BER}, we can always perform frequency alignment on any attacked signal, regardless of the attacks which are unknown beforehand. But frequency alignment probably would fail to extract the watermarks modified by other attacks rather than TSM/PSM attacks. Eventually, if the watermarks extracted by all attempts are unrecognized, the suspected signal is claimed to be unwatermarked. Therefore, the proposed audio watermarking scheme is undoubtedly blind.

Figure 4.12 shows the process of frequency alignment, where α is a scale factor, F_t is the frequency spectrum of the tth windowed frame in Eq. (4.9), and D_t is the resulting vector after performing frequency alignment. Note that we consider both positive- and negative-frequency parts of F_t during the calculation; hence, the length of F_t is N. However, only the first $\frac{N}{2}$ points of D_t, i.e., $D_t\left(1:\frac{N}{2}\right)$, will be taken as the descaled frequency spectra for further processing.

In general, the elements of F_t are indexed by a vector i_F, i.e., $i_F = 1 : N$. Another index vector i_X is calculated by rounding the result of i_F/α to the nearest integer, i.e., $i_X = \mathrm{round}\,(i_F/\alpha)$. The values of i_X in turn determines the resulting vector, D_t.

Specifically, for a positive PSM, $\alpha > 1$ may result in repetitive values of i_X, i.e., the nth, $(n + 1)$th, $\ldots$, and $(n + x)$th elements have the same value. If the nth element of i_X has a unique value m, then transfer $F_t(n)$ to $D_t(m)$. For instance, in our example in Fig. 4.12a, the first element of i_X has a unique value of 1, this indicates transferring $F_t(1)$ to $D_t(1)$. Otherwise, calculate the average of the nth to $(n + x)$th elements (which correspond to repetitive values of i_X) and transfer this value to $D_t(m)$. For instance, in our example, the third and fourth elements of i_X both have values of 3, which indicates transferring the average of $F_t(3)$ and $F_t(4)$ to $D_t(3)$.

As for negative PSM ($\alpha < 1$), the values of i_X are discontinuous. If the nth element of i_X has a value of n, then transfer $F_t(n)$ to $D_t(n)$. For instance, in our example in Fig. 4.12b, the first element of i_X equals 1, this indicates transferring $F_t(1)$ to $D_t(1)$. If the nth element of i_X has a value m, where $m \neq n$, then transfer $F_t(n)$ to $D_t(m)$, e.g., in our example, the fourth element of i_X equals 5, this indicates transferring $F_t(4)$ to $D_t(5)$. As the values of i_X is discontinuous, only part of the vector D_t is filled by this mechanism. The rest of the vector is calculated by linear interpolation between successive known values.

As shown above, the outcomes of the frequency alignment for a positive and negative PSM are different—one being compression and the other being expansion. Using this knowledge, we can use one positive and one negative trial values of the scale factor respectively to descale the frequency spectra. Two trial values are generally within $\pm 10\%$, e.g., $+6\%$ and -6%. After the frequency spectra are descaled, adaptive synchronization is performed to output the average synchronization strength A_{sync}. Let us denote the value of A_{sync} obtained from the positive trial value by A_{sync}^+ and that from the negative trial value by A_{sync}^-. If $A_{\text{sync}}^+ > A_{\text{sync}}^-$, it can be deduced that the watermarked signal has been attacked by a positive PSM, and vice versa. Further, the scale factor can be delicately adjusted for a higher detection rate.

To convert PITSM to TPPSM, the length of the host signal (N_o) is required to be the information shared between the embedder and the detector. By comparing with the length of the attacked signal (N_a), it is ascertained whether the attack is a positive or negative PITSM. Accordingly, the attacked signal is resampled to the corresponding TPPSM. Then, frequency alignment is performed to improve the accuracy of watermark detection. Although a slight deviation of N_a might occur (which happens when samples cropping or inserting attack the watermarked signal along with PITSM), experimental results in the next chapter show that such an amount of difference is negligible to the operation of resampling.

4.3.3.3 Error Analysis Associated with T_{BER}

As mentioned in Sect. 4.3.3.2, the BER threshold T_{BER} is used to determine whether one detection is successful or not and further to declare whether a watermark exists or not. In practice, we are given a signal under inspection and then perform watermark detection to extract a supposed watermark. Once one extracted watermark has a BER less or equal than T_{BER}, the suspected signal is claimed to be watermarked.

Otherwise, the suspected signal is claimed to be unwatermarked if all attempts to detect the watermark always get BERs more than T_{BER}.

Consequently, there are two types of errors in determining the existence of a watermark, i.e., a false-positive error and a false-negative error. A false-positive error (or false alarm) occurs when the detector indicates the presence of a watermark in an unwatermarked signal, and the false- positive probability is the likelihood of such an occurrence. A false- negative error (or miss detection) occurs when the detector indicates the absence of a watermark in a watermarked signal, and the false-negative probability is the likelihood of such an occurrence [13]. It is rather difficult to establish exact models for both false-positive and false-negative probabilities. Here, simple binomial models [2, 5, 130] are used in our analysis: the extracted N_w bits are assumed to be N_w independent, identically distributed Bernoulli variables with the same "success" probability p, where N_w is the watermark length. Note that a "success" means the extracted bit matching the original watermark bit.

According to the definition, a false-positive error of one detection occurs if this detection extracts $(N_w - N_e)$ or more bits successfully, where $N_e = \lfloor N_w \times T_{\text{BER}} \rfloor$ is the number of wrong bits. Since the "success" probability p of each bit extracted from an unwatermarked signal is supposed to be $\frac{1}{2}$, the false-positive probability of one detection P_{pd} is calculated as follows [2, 5, 130]:

$$
\begin{aligned}
P_{pd} &= \sum_{k=(N_w-N_e)}^{N_w} C\left(N_w, k\right) \left(\frac{1}{2}\right)^k \left(1 - \frac{1}{2}\right)^{(N_w-k)} \\
&= \frac{1}{2^{N_w}} \sum_{k=(N_w-N_e)}^{N_w} C\left(N_w, k\right),
\end{aligned}
\tag{4.28}
$$

where $C\left(N_w, n\right)$ denotes the number of combinations, i.e., $C\left(N_w, n\right) = \frac{N_w!}{n!(N_w-n)!}$. Furthermore, since the suspected signal is claimed to be watermarked as soon as one detection is successful, the false-positive probability of watermark existence is determined by [2, 130]

$$
P_{pw} = \sum_{n=1}^{N_{\text{det}}} C\left(N_{\text{det}}, n\right) \left(P_{pd}\right)^n \left(1 - P_{pd}\right)^{(N_{\text{det}}-n)},
\tag{4.29}
$$

where N_{det} is the total number of detections that were performed.

The analysis of false-negative probability is more complicated than false-positive probability, because false-positive probability depends on the watermark detection algorithm only; however, false-negative probability depends on both watermark embedding and detection algorithms. Moreover, since the watermarked signal might be distorted by various attacks, an accurate false-negative probability should be calculated for a specific attacked signal [13]. In our case, a general model is employed to roughly estimate the false-negative probability.

The false-negative probability is computed in a different way from the false-positive probability, i.e., calculating the correct detection probability which is merely the complement of the false-negative probability. Considering that the "success" probability p of each bit extracted from an attacked signal is approximated to be $(1 - T_{\mathrm{BER}})$, the correct detection probability of one detection P_{cd} is calculated as follows [130]:

$$P_{cd} = \sum_{k=(N_w-N_e)}^{N_w} C\,(N_w,\,k)\,(1 - T_{\mathrm{BER}})^k\,(T_{\mathrm{BER}})^{(N_w-k)}. \tag{4.30}$$

Recall that the suspected signal is claimed to be watermarked if at least one detection is successful. Therefore, the false-negative probability of watermark existence is determined by [130]

$$P_{nw} = 1 - \sum_{n=1}^{N_{\mathrm{det}}} C\,(N_{\mathrm{det}},\,n)\,(P_{cd})^n\,(1 - P_{cd})^{(N_{\mathrm{det}}-n)}. \tag{4.31}$$

The severity of false-positive and false-negative probabilities is application dependent. Mostly, more concern is focused on minimizing the occurrence of false-positive errors. For the application of copyrights protection, a very low false-positive probability (less than 10^{-5}) is of a higher priority [5, 130], and meanwhile a moderate false-negative probability is desired under various attacks within tolerable levels.

4.4 Coded-Image Watermark

From the analysis of the embedding and detection algorithms, the watermark embedded for copyrights protection is essentially represented by a series of watermark bits.

To better function for copyrights protection, the proposed scheme adopts coded-image such as WATERMARK with bit "1" and "0" (mapped to "−1")[5] as a visual watermark, instead of a meaningless pseudorandom or chaotic sequence. A coded-image can be identified visually, as a kind of ownership stamp. Moreover, post-processing on the extracted watermark can also be done to enhance the binary image and consequently the detection accuracy will increase. Image denoising and pattern recognition are examples of post-processing techniques for automatic character recognition. Thus, on top of BER, coded-image provides a semantic meaning for reliable verification [131]. In addition, as mentioned in Sect. 4.2.2, the watermarking scheme benefits from the encryption of the coded-image to obtain extra security.

---

[5]By definition, a coded-image belongs to a binary image, which has only two values for each pixel.

Fig. 4.13 Coded-image
denoising by morphological
operations

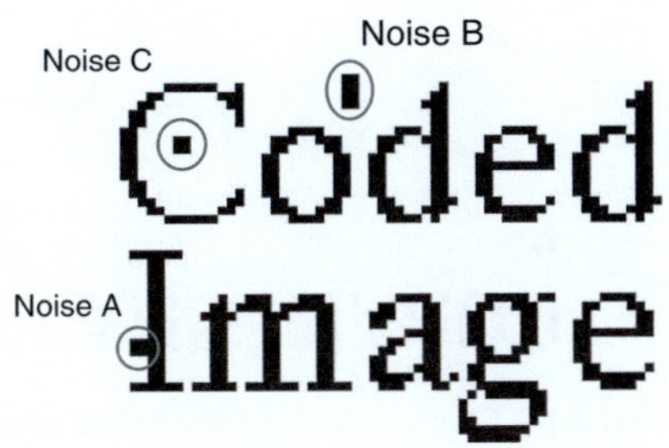

In our scheme, a coded-image watermark consists of letters, each of which is represented by a matrix of 7×5 bits. It is worth mentioning that the coded-image watermark is embedded letter by letter, not line by line. In this way, even when a part of the watermarked signal is severely attacked, the letters in the other parts still can be clearly recognized.

Discussion of post-processing techniques on binary images is beyond the scope of this thesis. For completeness, two possible methods are briefly described below, i.e., morphology and neural network (NN) [132].

Morphology is a technique of image processing based on shapes. The fundamental morphological operations are erosion and dilation, which can be used in a variety of ways to give other transformations including opening, closing, skeletonization, and so on. A simple application of the opening operation is to remove small objects with fewer pixels than a threshold from the image. For example, noise B and C in Fig. 4.13 can be easily eliminated by this means. However, it is impossible to get rid of noise A which is connected with the informative letters. In this case, skeleton-based character recognition might be a good solution to retrieve the distorted coded-image. But character skeletonization is not quite applicable to the proposed scheme, since the line width of the coded-image does not contain a large number of pixels.

Neural networks have been widely used for character recognition. In this book, a backpropagation neural network from the Neural Network Toolbox for MATLAB[6] is employed to recognize all 26 capital letters of the alphabet. Each letter is represented as a 7 by 5 matrix, such as the letters shown in Fig. 4.14a. As illustrated in Fig. 4.14b, c, noisy letters on the extracted coded-image watermark could be fully recovered by the neural network.

---

[6]Appcr1: Character Recognition at http://www.mathworks.com/access/helpdesk/help/toolbox/nnet/.

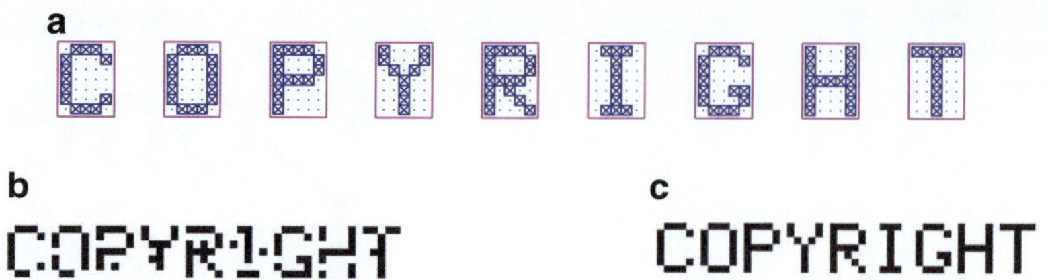

Fig. 4.14 Character recognition by the neural network. (**a**) Letters "C" "O" "P" "Y" "R" "I" "G" "H" "T". (**b**) Noisy coded-image watermark. (**c**) Recovered coded-image watermark

4.5 Summary

As discussed in Chap. 3, spread spectrum technique contributes to the robustness and security of audio watermarking schemes. However, traditional SS watermarking is likely to suffer from perceivable watermark embedding and desynchronization attacks. In this chapter, an imperceptible, robust, and secure audio watermarking scheme using Psychoacoustic Model 1, multiple scrambling, adaptive synchronization, frequency alignment, and the coded-image watermark is proposed. Essentially, a watermark modulated by PRNs is embedded in the time–frequency domain of the host signal. Accordingly, watermark detection is based on normalized correlation between the watermarked signal and corresponding PRNs.

To overcome the problem of perceivable watermark embedding, only the regions with high energy are selected to prevent noise caused by watermarking the silence. Moreover, a half-overlapped Hanning window and the GTF involved in the construction of the watermarking domain are also beneficial to transparent perception. Most significantly, the MMT from Psychoacoustic Model 1 is utilized to shape the amplitude of the watermark signal, so that the watermarks are embedded without noticeably degrading the perceptual quality of the audio signals.

Meanwhile, several measures have been taken to cope with various attacks, especially desynchronization attacks. Commonly, basic detection with block synchronization is applied to seek the positions of synchronization bits as well as watermark bits. However, the synchronization positions found are considered incorrect when exceeding a certain threshold. In such cases, adaptive synchronization is employed to search for the correct synchronization positions. Furthermore, to resist severe PITSM and TPPSM of up to $\pm 10\%$, frequency alignment is developed to descale the distorted frequency spectra of the attacked watermarked signal. Thus, the synchronization positions can be retrieved for successful watermark detection.

On top of imperceptibility and robustness, the proposed scheme is also strictly self-secured by using multiple scrambling. It is extremely difficult for any attacker lacking all the secret keys to ascertain or destroy the embedded watermark. In addition to the security benefited from the encryption, the usage of the coded-image watermark as a visual identification makes it possible to improve the accuracy of watermark detection further by employing image processing techniques and pattern matching analysis.

This chapter has presented a theoretical analysis of the embedding and detection algorithms. In Chap. 5, the performance of the proposed audio watermarking scheme will be thoroughly evaluated through perceptual quality assessment, robustness test and security analysis, etc.

Chapter 5
Performance Evaluation of Audio Watermarking

In Chap. 4, the embedding and detection algorithms of the proposed audio watermarking scheme were analyzed theoretically. The aim of this chapter is to examine system performance in terms of imperceptibility, robustness, security, data payload, and computational complexity, as required in Sect. 1.3.1.

First, the process of determining the parameters used for watermarking is described. Then performance measurement begins with perceptual quality assessment, which consists of the subjective listening test and the objective evaluation test. This is subsequently followed by a complete robustness test including both basic and advanced robustness tests. After performing a security analysis, we carry out the estimations of data payload and computational complexity. Finally, a performance comparison is made between the proposed scheme and other existing schemes. Some observations are discussed according to the experimental results.

5.1 Experimental Setup

To investigate the performance of the proposed audio watermarking scheme, a series of experiments were carried out on different audio signals. It is worth mentioning that all the audio signals in the test set are taken as host audio signals in order to inspect the applicability of the proposed scheme. But for the sake of illustration, we choose one typical audio signal from each category only and present their simulation results separately in the following sections. Nevertheless, similar results can be found for other signals. The selected audio files are (i) vocal, *Bass.wav* (A_2); (ii) percussive instruments, *Glockenspiel2.wav* (A_7); (iii) tonal instruments, *Harpsichord.wav* (A_8); and (iv) music, *Pop.wav* (A_{15}). For simplicity, *Glockenspiel2.wav* and *Harpsichord.wav* are shortened as *Gspi.wav* and *Harp.wav* respectively.

Recall that audio watermarking always involves the trade-off relationships among imperceptibility, robustness, security, data payload, computational complexity, and so on. Thus, experiment parameters should be properly set to optimize the

Y. Lin and W.H. Abdulla, *Audio Watermark: A Comprehensive Foundation Using MATLAB*, 123
DOI 10.1007/978-3-319-07974-5__5, © Springer International Publishing Switzerland 2015

system performance for the intended application. As mentioned in Sect. 1.2.3.1, imperceptibility, robustness, and security are the key criteria in designing any audio watermarking scheme for copyrights protection. Accordingly, the parameters involved in the watermarking are determined based on their requirements.

- Determination of the variables N, N_{unit}, $N_{subband}$, N_{bit}, and N_B

The variables associated with the structure of the watermarking domain defined in Sect. 4.1.2 are first identified, including the frame length (N), the number of units per block (N_{unit}), the number of nonuniform subbands ($N_{subband}$), the number of watermark bits embedded in one block (N_{bit}), and the number of slots for embedding each watermark bit (N_B).

Note that the value of frame length is fixed at $N = 512$ due to the use of Psychoacoustic Model 1, and hence $\frac{N}{2} = \frac{512}{2} = 256$ FFT coefficients per frame are available for watermarking. Moreover, as mentioned in Sect. 4.1.3, each tile is required to contain more than five FFT coefficients for the purpose of robustness. Following the calculations for the channels of the Gammatone filterbank, a proper value of $N_{subband}$ is obtained as $N_{subband} = 32$. On considering the operation of multiple scrambling, the number of selected subbands $\tilde{N}_{subband}$ is set to be 28. Then, the values of N_{bit} and N_B are subsequently decided. So with N_{unit}, $N_{subband}$, N_{bit}, and N_B, the number of slots for embedding the synchronization bit (N_s) can be calculated by Eq. (4.3) as shown below:

$$N_s = N_{unit} \times \tilde{N}_{subband} - N_{bit} \times N_B \tag{5.1}$$

These variables combine to affect the system performance in some way. According to the embedding algorithm previously described, larger N_s and N_B could contribute to a stronger robustness against desynchronization attacks to some extent. Correspondingly, a larger value of N_{unit} and a smaller value of N_{bit} are desired. However, a larger N_{unit} would lead to an increase in computational complexity. As indicated in Eq. (4.18), the number of times that each block searches for its synchronization position is equal to $N_f = N_c \times N_{unit}$, where $N_c = 4$. Also, the values of all tiles in the blocks should be provided simultaneously for watermark detection, thus more computer memory is required to store the data. Additionally, as a result of a smaller N_{bit}, data payload would be inevitably reduced.

In view of these constraints, the above variables are specified as follows to aim for a good compromise between various requirements. That is, $N_{unit} = 10$, $N_{bit} = 4$, $N_B = 30$, and the resulted $N_s = 160$, which will be employed as constants for all experiments.

- Determination of watermark strength α_w

The amplitude of the watermark signal in the embedding has an important influence on system performance. As discussed in Sect. 4.2.1, the magnitude spectrum of the watermark signal is controlled by the watermark strength α_w. To get good performance in the experiments, watermark strength might be selected to be uniformly distributed between 10 and 200, i.e., $\alpha_w = 10, 20, 30, \ldots, 200$.

Given a host audio signal, watermarking with a smaller α_w would result in better imperceptibility, but weaker robustness; on the other hand, watermarking with a larger α_w would result in imperceptibility degradation, but stronger robustness [19]. Considering that imperceptibility is a prerequisite for practical application of audio watermarking, our prime concern in determining α_w is dedicated to maintaining the perceptual quality of the watermarked audio signals. Within the scope of satisfactory perceptual quality, a value of α_w that provides adequate robustness is adopted. In our experiments, the software "Perceptual Evaluation of Audio Quality" (PEAQ) [48] is used to interpret the perceptual quality, and an objective difference grade (ODG) within $[-2.0, 0]$ is deemed to be acceptable. Also, the property of robustness is denoted by the bit error rate (BER) of the extracted watermark under 36 dB noise addition attack. [1] Empirically, a reasonable BER is expected to be less than 10 %.

For example, Fig. 5.1 shows the determination of watermark strength α_w for *Bass.wav* and *Pop.wav*. As indicated in Fig. 5.1a, watermarking with $\alpha_w \leq 50$ is considered unperceived, since the ODGs fit within the allowable range $[-2.0, 0]$. Also, under the condition of $\alpha_w \geq 50$, the BERs are less than 10 % and thereby the requirement of robustness is met. Consequently, $\alpha_w = 50$ is the only appropriate value for watermarking *Bass.wav*. By comparison, embedding a robust watermark into *Pop.wav* while retaining the imperceptibility is more feasible. With the same method, it is found that a proper watermark strength for watermarking *Pop.wav* ranges between 60 and 140, i.e., $60 \leq \alpha_w \leq 140$. In this case, we use the average value $\alpha_w = 100$ for the experiments with *Pop.wav* below.

- Determination of the embedded watermark

Generally, the less watermark bits embedded, the better imperceptibility but the worse robustness. To evaluate the proposed scheme fairly, our experiments always embed the watermark bits into host signals at full capacity. For example, the number of watermark bits (N_w) embedded into *Bass.wav*, *Gspi.wav*, *Harp.wav*, and *Pop.wav* is 350, 210, 140, and 280, respectively. More analysis of data payload will be presented later in Sect. 5.5.1.

As shown in Sect. 4.4, a coded-image watermark offers greater advantage over a pseudorandom sequence (PRS). Therefore, the coded-image watermark is always adopted in the experiments. Recall that each letter on the coded-image watermark is represented by a matrix of 7×5 bits, i.e., $L_w = 35$ bits for one letter. Thus, N_w watermark bits can be coded into $N_L = N_w/L_w$ letters. For example, the coded-image watermark embedded into *Bass.wav* is PROTECTION, which consists of $N_L = 350/35 = 10$ letters; the coded-image watermark embedded into *Gspi.wav* is ROBUST, which consists of $N_L = 210/35 = 6$ letters; the coded-image watermark

---

[1]Additive noise attack is a commonly used attack in robustness test of audio watermarking techniques. As clearly indicated in Appendix A and B, SDMI standard and STEP 2000 employ 36 dB and 40 dB additive noise attack respectively. Therefore, a rigorous additive noise attack with a lower SNR value, i.e., 36 dB additive noise attack, is chosen for our basic robustness test listed in Appendix E.

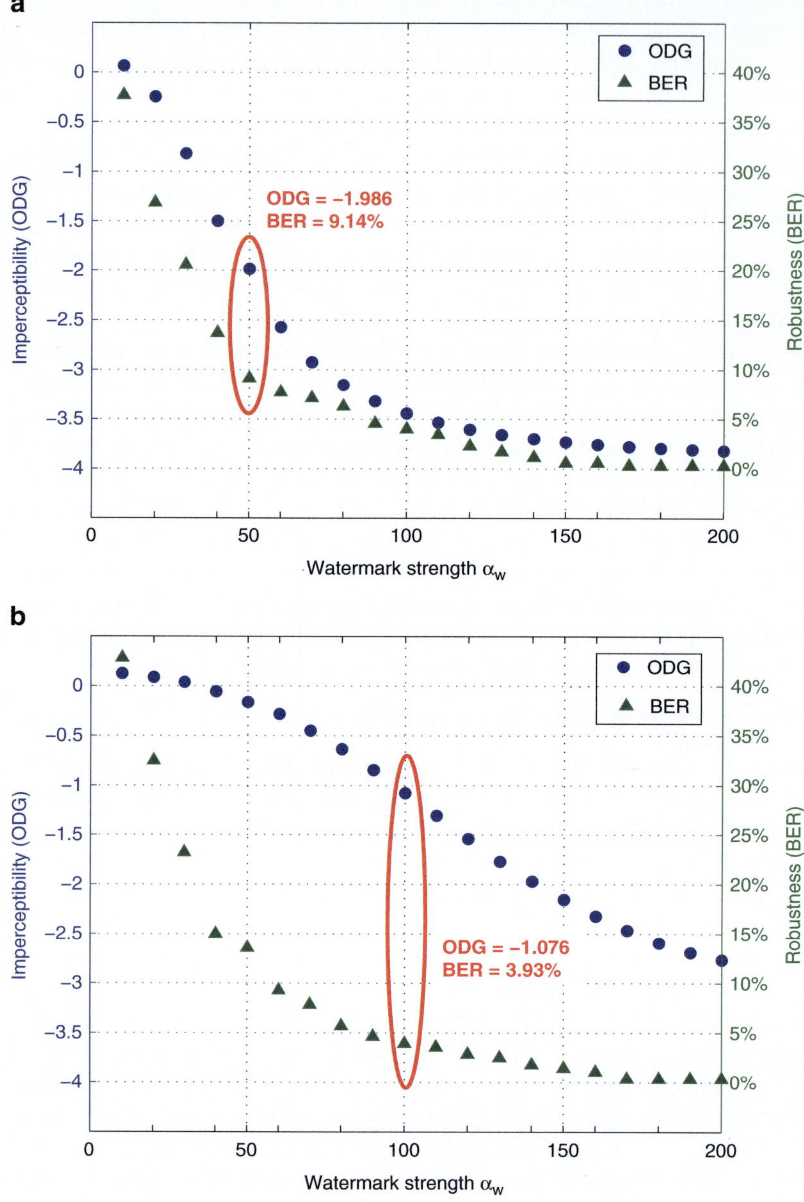

Fig. 5.1 Determination of watermark strength. α_w (**a**) $\alpha_w = 50$ for *Bass.wav*. (**b**) $\alpha_w = 100$ for *Pop.wav*

embedded into *Harp.wav* is MARK, which consists of $N_L = 140/35 = 4$ letters; and the coded-image watermark embedded into *Pop.wav* is SECURITY, which consists of $N_L = 280/35 = 8$ letters.

Based on the above parameters, the audio signals in the test set are watermarked. Then, the watermarked *Bass.wav*, *Gspi.wav*, *Harp.wav*, and *Pop.wav* signals are used for illustration in the following sections, except for the test under collusion in Sect. 5.3.3.2 and the test under multiple watermarking in Sect. 5.3.3.3. In the tests under collusion and multiple watermarking, a number of coded-image watermarks are embedded into each host signal as required.

5.2 Perceptual Quality Assessment

The goal of perceptual quality assessment is to fairly judge the perceptual quality of the watermarked audio signals relative to host audio signals. To this end, both subjective and objective approaches to perceptual quality assessment are employed in this book, as discussed in Sect. 1.3.2.1.

5.2.1 Subjective Listening Test

Subjective listening tests are carried out in two ways: the MUSHRA test and the five-scale subjective difference grade (SDG) rating. As described in Sect. 3.1.2, listening tests performed in an isolated chamber were undertaken by ten trained participants and all the stimuli are presented through a high-fidelity headphone.

The MUSHRA test stands for MUlti Stimuli with Hidden Reference and Anchors test, which is defined by ITU-R recommendation BS.1534 [44]. In the MUSHRA test, the participant is exposed to three types of audio clips as test, reference (i.e., the original unprocessed audio), and anchor audio signals. The recommendation specifies that one anchor must be a 3.5 kHz low-pass filtered version of the reference audio signal [44]. Also, a hidden reference is usually adopted as another anchor. Then, the participant is asked to grade the perceptual quality of the audio signals under test and the anchors relative to the reference audio signal.

We developed a MATLAB GUI for the MUSHRA test to help our analysis, as shown in Fig. 5.2. In the context of audio watermarking, the watermarked signal is the signal under test, while the host signal is the reference signal. As required, the host signal is always presented in the experiments. For the anchors, we use three versions of the host signal, i.e., a hidden version, a 3.5 kHz low-pass filtered version,

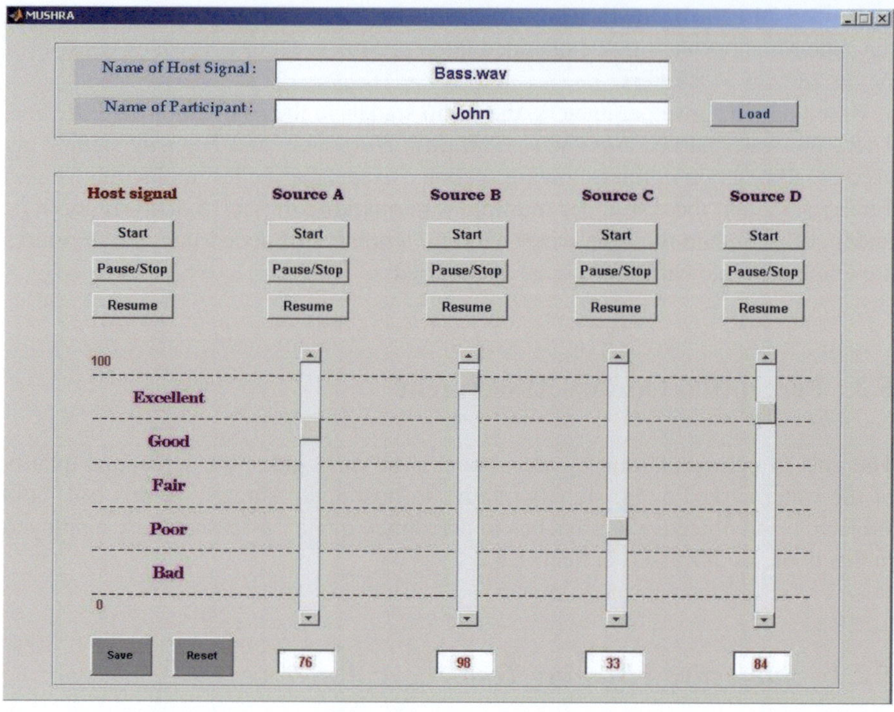

Fig. 5.2 Screenshot of the MATLAB GUI for the MUSHRA test. The buttons on the GUI have the following functions: "Load," load the host audio signal to be evaluated. "Start," start playing the sound from the beginning. "Pause/Stop," pause or stop the sound that is currently playing. "Resume," resume the sound from the pause position. "Save," save the host signal name and the participant name as well as the registered scores into a .txt file. "Reset," reset the interface for the next trial

and a 96 kbps MP3 compressed version.[2] During each experiment, the participant is therefore asked to grade four versions of a given host signal, i.e., the watermarked signal (WM), the hidden reference (HOST), the low-pass filtered version (LPF), and the compressed version (MP3).

Given a host signal, one participant can launch the MUSHRA test by clicking the "Load" button. Subsequently, four versions of the host signal will be randomly assigned to Source A~D. Then, the participant grades each version by moving the slider to the location corresponding to the perception. Accordingly, a score between [0, 100] appears in the text box below. With the buttons "Start," "Pause/Stop," and "Resume," the participant can switch instantly between different sound files. Finally, the buttons "Save" and "Reset" save the test results and get ready for next experiment.

---

[2]The 3.5 kHz low-pass filtered version refers to a version of host audio filtered by a 3.5 kHz low-pass filter, and the 96 kbps MP3 compressed version refers to a version of host audio after MP3 compression at 96 kbps.

As mentioned in Sect. 3.1.1, the test set includes 17 pieces of audio signals, denoted by A_i, $i = 1, 2, \ldots, 17$. Then, the four versions of A_i are denoted by A_{ij}, where $j = 1, 2, 3,$ and 4 stands for the version of WM, HOST, LPF, and MP3, respectively. Since there are 10 subjects participating in the tests, the score of A_{ij} provided by the k-th subject is denoted by $G_M(i, j, k)$, where $k = 1, 2, \ldots, 10$.

After all the scores are collected, statistical analysis [44] is performed to assess the perceptual quality of each A_{ij} separately. First, the mean of the scores of A_{ij} is calculated by

$$\overline{\mu}_{ij} = \frac{1}{K} \sum_{k=1}^{K} G_M(i, j, k) \tag{5.2}$$

where $K = 10$ in our experiments.

Then, a 95 % confidence interval ($\alpha = 1 - 0.95 = 0.05$) about the mean value $\overline{\mu}_{ij}$ is given by

$$\left[\overline{\mu}_{ij} - \delta_{ij}, \ \overline{\mu}_{ij} + \delta_{ij} \right] \tag{5.3}$$

where

$$\delta_{ij} = t_{0.05} \frac{\sigma_{ij}}{\sqrt{K}} \tag{5.4}$$

Here, $t_{0.05}$ is the t test for a significance level of 95 % and σ is the standard deviation defined as

$$\sigma_{ij} = \sqrt{\frac{1}{(K-1)} \sum_{k=1}^{K} \left[G_M(i, j, k) - \overline{\mu}_{ij} \right]^2} \tag{5.5}$$

On the assumption that the mean scores follow a normal distribution, the value of $t_{0.05}$ is equal to

$$t_{0.05} = \Phi^{-1}\left(1 - \frac{\alpha}{2}\right) = \Phi^{-1}(0.975) = 1.96 \tag{5.6}$$

where $\Phi^{-1}(\cdot)$ is the inverse normal cumulative distribution function.

After substituting $t_{0.05}$ in Eq. (5.6) into Eq. (5.4), the 95 % confidence interval in Eq. (5.3) becomes

$$\left[\overline{\mu}_{ij} - 0.62\sigma_{ij}, \ \overline{\mu}_{ij} + 0.62\sigma_{ij} \right] \tag{5.7}$$

Figure 5.3 shows the results of statistical analysis on *Bass.wav*, *Gspi.wav*, *Harp.wav*, and *Pop.wav*. Different versions of each host signal are spread along the

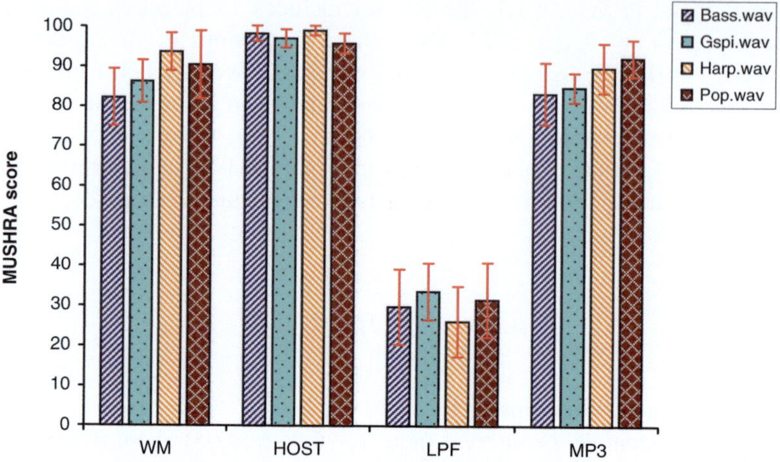

Fig. 5.3 Results of the MUSHRA-based subjective listening test

X-axis, while the perceptual quality scores are lying along the Y-axis. The average scores over ten listeners as well as the related 95 % confidence interval are displayed.

From Fig. 5.3, the hidden references are rated the highest with small 95 % confidence intervals. On the other hand, the scores of 3.5 kHz low-pass filtered signals are quite low. This is because a low-pass filter attenuates high frequencies, which makes audio samples sound dull. Moreover, the scores of watermarked signals are comparable to those of MP3-compressed signals at 96 kbps, which means they are of similar perceptual quality. For different host audio signals, the watermarked *Harp.wav* signal has the best performance and its average score is around 93. The second is the watermarked *Pop.wav* signal, followed by the watermarked *Gspi.wav* signal. Although the watermarked *Bass.wav* signal obtains the lowest average score, the score is still more than 80. Therefore, we conclude that perceptual quality of all the watermarked signals is well preserved. Also, perceptual quality depends on the music type. It is worth mentioning that three observations are common issues among all techniques, due to the complexity of audio signals [133].

In addition to MUSHRA test, a rating based on the five-scale SDG evaluates the perceptual quality of the watermarked signals in a straightforward manner. The subjects are asked to rate a watermarked signal relative to its host signal according to the descriptions in Table 1.2. Similarly, the SDG of host signal A_i from the k-th subject is denoted as $G_{SDG}(i, k)$, where $i = 1, 2, \ldots, 17$ and $k = 1, 2, \ldots, 10$. Then, the average SDG of host signal A_i is calculated as

$$\bar{v}_i = \frac{1}{K} \sum_{k=1}^{K} G_{SDG}(i, k) \tag{5.8}$$

where $K = 10$.

Table 5.1 Results of the SDG-based subjective listening test

	Bass.wav	*Gspi.wav*	*Harp.wav*	*Pop.wav*
Average SDG	-1.52	-1.27	-0.45	-0.83

Table 5.2 Results of the objective evaluation test

	Bass.wav	*Gspi.wav*	*Harp.wav*	*Pop.wav*
ODG	-1.986	-1.758	-0.509	-1.076
SNR/dB	33.39	32.01	24.44	30.43

For example, the average SDGs for the watermarked *Bass.wav*, *Gspi.wav*, *Harp.wav*, and *Pop.wav* signals are shown in Table 5.1. From the table, it is seen that the average SDGs for these samples are in the range of -1.6 and 0. In fact, except that a few feel a slight difference between *Bass.wav* and its watermarked signal, most listeners find it hard to distinguish the host and the watermarked audio signals during the experiments.

5.2.2 Objective Evaluation Test

As discussed in Sect. 3.1.2, objective evaluation tests on the watermarked audio signals include two metrics, namely the ODG by using software PEAQ and the signal-to-noise ratio (SNR) defined in eq. (1.3).

The values of the ODG and the SNR for the watermarked *Bass.wav*, *Gspi.wav*, *Harp.wav*, and *Pop.wav* signals are shown in Table 5.2. As can be seen, it is acceptable that the ODGs are within $(-2.0, 0)$ below the threshold of becoming slightly annoying. By averaging these four values, the average ODG of the proposed scheme is calculated to be -1.33. Moreover, the ODG values accord with the results of the subjective listening test. That is, the watermarked *Harp.wav* signal is the most imperceptible, followed in descending order by the watermarked *Pop.wav*, *Gspi.wav*, and *Bass.wav* signals.

In addition, the SNR values are higher than the 20 dB requirement from the International Federation of the Phonographic Industry (IFPI) [2, 14]. By averaging these four values, the average SNR value of the proposed scheme is calculated to be 30.1 dB.

From Tables 5.1 and 5.2, one point to note is that the SNR values are not in agreement with the actual perceptual quality in terms of the SDG. More investigation on other objective quality measures will be made in the next chapter.

From the above results of perceptual quality assessment, it is verified that our watermarked signals are mostly perceptually undistinguished from the host audio signals.

5.3 Robustness Test

The goal of the robustness test is to investigate the capability of the watermarked audio signals to resist various attacks. To fully evaluate the robustness of the proposed audio watermarking scheme, we carry out both basic and advanced robustness tests as depicted in Sect. 3.1.3. All ODGs in this section are provided by using PEAQ.

5.3.1 Error Probability

Recall from Sect. 4.3.3.3 that the threshold T_{BER} has an influence on both false-positive (P_{pw}) and false-negative (P_{nw}) probabilities of declaring the existence of a watermark. Suppose that the value of T_{BER} is equal to 20 % [2,5], P_{pw} and P_{nw} are calculated using Eqs. (4.28)–(4.31).

Table 5.3 shows the results on the error probabilities of the watermarked *Bass.wav*, *Gspi.wav*, *Harp.wav*, and *Pop.wav* signals, where N_w is the watermark length and $N_e = \lfloor N_w \times T_{BER} \rfloor$ is the number of wrong bits. As mentioned in Sect. 5.1, 350, 210, 140, and 280 watermark bits are separately embedded into *Bass.wav*, *Gspi.wav*, *Harp.wav*, and *Pop.wav*, so the resulting N_e is equal to 70, 42, 28, and 56, respectively. In addition, N_{det} is the number of detections performed, where two values (i.e., $N_{det} = 5$ and 10) are considered.

For different host signals in Table 5.3, the false-positive probabilities P_{pw} increase exponentially with the watermark length N_w, but vary slightly with the number of detections performed N_{det}. On the other hand, the false-negative probabilities P_{nw} increase with N_{det}, but vary slightly with N_w.

Generally, the severity of false-positive and false-negative probabilities is application dependent. In our scheme, given $T_{BER} = 20\%$, the false-positive probabilities have already satisfied the requirement, being much less than 10^{-5}. When $N_{det} = 5$ and 10, the false- negative probabilities are around 10^{-2} and 10^{-4}, respectively. These values are sufficient for the application of copyrights protection [130]. Therefore, the threshold is set to be $T_{BER} = 20\%$ in our experiments. This means that the detections with the BERs of greater than 20 % are considered failed.

Table 5.3 Results of error probabilities under $T_{BER} = 20\%$

		Bass.wav	Gspi.wav	Harp.wav	Pop.wav
	N_w	350	210	140	280
	N_e	70	42	28	56
$N_{det} = 5$	P_{pw}	1.78×10^{-30}	1.20×10^{-18}	1.06×10^{-12}	1.44×10^{-24}
	P_{nw}	2.25×10^{-2}	2.03×10^{-2}	1.84×10^{-2}	2.16×10^{-2}
$N_{det} = 10$	P_{pw}	3.57×10^{-30}	2.40×10^{-18}	2.12×10^{-12}	2.88×10^{-24}
	P_{nw}	5.05×10^{-4}	4.14×10^{-4}	3.38×10^{-4}	4.66×10^{-4}

5.3.2 Basic Robustness Test

As described in Sect. 3.1.3.1, a variety of common signal operations and desynchronization attacks are included in the basic robustness test, for example, noise addition, resampling, requantization, amplitude scaling, low-pass filtering, echo addition, reverberation, MP3 compression, DA/AD conversion, random samples cropping, jittering, zeros inserting, pitch-invariant time-scale modification (PITSM), and tempo-preserved pitch-scale modification (TPPSM). Moreover, a number of combined attacks are constructed to test the robustness, such as random samples cropping, jittering, or zeros inserting followed by low-pass filtering or MP3 compression. Also, PITSM or TPPSM followed by low-pass filtering, MP3 compression, random samples cropping, jittering, or zeros inserting are employed to attack the watermarked signals, in order to verify the validity of frequency alignment under challenging conditions.

As described in Sect. 4.3, the whole procedure of watermark detection begins with basic detection. Then, basic detection updated with adaptive synchronization (called improved detection, for simplicity) is always performed to improve the detection rate. In cases of PITSM and TPPSM, frequency alignment is further used to descale the frequency spectra with an attempt to retrieve the distorted watermark. For simplicity, improved detection integrated with frequency alignment is called advanced detection. Note that it is not always necessary to proceed with frequency alignment to combat PITSM and TPPSM attacks. Empirically, if the improved detection has extracted a watermark with a BER of less than 10 %, such a result is considered to be already satisfactory for detection.

Tables 5.4, 5.5, 5.6, and 5.7 show the results of the basic robustness test on the watermarked *Bass.wav*, *Gspi.wav*, *Harp.wav*, and *Pop.wav* signals, respectively. Recall that the coded-image watermarks embedded are PROTECTION for *Bass.wav*, ROBUST for *Gspi.wav*, MARK for *Harp.wav*, and SECURITY for *Pop.wav*.

To illustrate the effectiveness of adaptive synchronization and frequency alignment, the BERs of the extracted watermarks at each stage of the detection are separately presented. Specifically, the "Basic detection" column, the "Adaptive synchronization" column, and the "Frequency alignment" column show the BERs of the extracted watermarks obtained by basic detection, improved detection, and advanced detection, respectively. Then, the extracted watermark obtained by the last detection is taken as the final extracted watermark, w_e, illustrated in the last column.

As discussed in Sect. 5.3.1, $T_{BER} = 20\%$. Accordingly, detections with BERs of greater than 20 % are considered failed and indicated by the symbol "×" in the tables. Moreover, the symbol "−" indicates the situation whereby advanced detection with frequency alignment is unexecuted. This is because the BER of the extracted watermark obtained by improved detection is already less than 10 %.

The following observations are obtained from the analysis of Tables 5.4, 5.5, 5.6, and 5.7.

- On the whole, the proposed audio watermarking scheme demonstrates strong robustness against various attacks, although different watermarked audio signals

Table 5.4 Results of the basic robustness test on the watermarked *Bass.wav* signal

		Basic detection	Adaptive synchronization	Frequency alignment	Final watermark
		(BER: %)			w_e
No attack		0	0	—	PROTECTION
SNR (dB)	40 dB	6.86	5.71	—	PROTECTION
	36 dB	11.71	9.14	—	PROTECTION
	30 dB	×	18.57	—	PROTECTION
Resampling (22.05 kHz)		0	0	—	PROTECTION
Requantization (8 bit)		17.71	16.00	—	PROTECTION
Amplitude	+20 %	0	0	—	PROTECTION
	−20 %	0	0	—	PROTECTION
Lp filtering	8 kHz	0	0	—	PROTECTION
	6 kHz	1.43	1.43	—	PROTECTION
	5 kHz	9.71	9.71	—	PROTECTION
DA/AD (line-in jack)		×	0	—	PROTECTION
Echo (0.3, 200 ms)		0.57	0.57	—	PROTECTION
Reverb (1 s)		0	0	—	PROTECTION
Compression II	96 kbps	0	0	—	PROTECTION
	64 kbps	1.43	1.14	—	PROTECTION
	48 kbps	3.71	2.86	—	PROTECTION
Cropping (8 × 25 ms)		×	0	—	PROTECTION
Jittering (0.1 ms/20 ms)		×	0	—	PROTECTION
Inserting (8 × 25 ms)		0	0	—	PROTECTION
PITSM	+4 %	×	0.57	—	PROTECTION
	+10 %	×	×	2.86	PROTECTION
	−4 %	×	1.71	—	PROTECTION
	−10 %	×	6.00	—	PROTECTION
TPPSM	+4 %	14.86	8.00	—	PROTECTION
	+10 %	×	×	4.86	PROTECTION
	−4 %	12.57	6.00	—	PROTECTION
	−10 %	×	×	4.29	PROTECTION

Notes: 1. Symbol "×": one detection with a BER of greater than 20 %
2. Symbol "−": one unexecuted advanced detection

differ in the performance. In terms of the BER, almost all the BERs of w_e are less than 10 %. Moreover, the coded-image watermarks embedded in the four host signals can always be extracted and clearly identified. Therefore, compared to merely meaningless bits, extra assistance in confirmation can be obtained from the coded-image watermarks. Although some pixels in the coded-image are mistaken, we are still able to recognize the copyrights information.

- For different watermarked signals, the watermarked *Harp.wav* signal in Table 5.6 shows the strongest robustness. None of the BERs are greater than 10 %, in fact most of them are equal to 0 %. This is because a higher watermark strength α_w is used in watermarking *Harp.wav*, as can be seen from the lower SNR value of the watermarked *Harp.wav* signal in Table 5.2. Due to the characteristics of the

Table 5.5 Results of the basic robustness test on the watermarked *Gspi.wav* signal

		Basic detection	Adaptive synchronization	Frequency alignment	Final watermark
		(BER: %)			w_e
No attack		0	0	—	ROBUST
SNR (dB)	40 dB	12.38	7.62	—	ROBUST
	36 dB	12.38	9.52	—	RUBUSJ
	30 dB	16.67	16.19	—	RƆBUSƷ
Resampling (22.05 kHz)		1.90	1.90	—	ROBUST.
Requantization (8 bit)		19.05	19.05	—	RUP.ISR
Amplitude	$+20\%$	0	0	—	ROBUST
	-20%	0	0	—	ROBUST
Lp filtering	8 kHz	0.95	0.95	—	ROBUST
	6 kHz	6.67	4.29	—	RƟBUSƬ
	5 kHz	×	11.43	—	RƆƷU3ʔ·
DA/AD (line-in jack)		×	0	—	ROBUST
Echo (0.3, 200 ms)		0	0	—	ROBUST
Reverb (1 s)		0	0	—	ROBUST
Compression II	96 kbps	2.86	2.86	—	ROBUSƬ
	64 kbps	3.81	3.81	—	ROBUSƬ
	48 kbps	14.76	9.05	—	ROƷUSƮ
Cropping (8 × 25 ms)		×	0	—	ROBUST
Jittering (0.1 ms/20 ms)		×	3.33	—	ROBUSƬ
Inserting (8 × 25 ms)		0	0	—	ROBUST
PITSM	$+4\%$	0.48	0.48	—	ROBUST
	$+10,\%$	×	×	10.00	RƟBUƐƬ
	-4%	×	4.29	—	RƟBUSƬ
	-10%	×	10.48	4.29	RƟBUST
TPPSM	$+4\%$	×	10.95	3.81	ROBUSƬ
	$+10\%$	×	×	9.52	RƟBUSƮ
	-4%	×	12.86	4.29	ROBÖST
	-10%	×	×	11.90	RƃBUSƬ

Notes: 1. Symbol "×": one detection with a BER of greater than 20 %
2. Symbol "—": one unexecuted advanced detection

harpsichord, *Harp.wav* can be watermarked with a higher α_w on the premise of a satisfactory perceptual quality.

This is followed by the watermarked *Bass.wav* signal in Table 5.4 and the watermarked *Pop.wav* signal in Table 5.7. For the watermarked *Bass.wav* signal, all the BERs of w_e are less than 10 %, except for the ones under 30 dB noise addition and requantization attacks. Similarly, for the watermarked *Pop.wav* signal, all the BERs of w_e are also less than 10 %, except for the ones under 30 dB noise addition, requantization, and $+10\%$ TPPSM attacks. On average, the watermarked *Gspi.wav* signal in Table 5.5 shows the weakest resistance to the attacks. Four final watermarks, which are attacked by 30 dB noise addition,

Table 5.6 Results of the basic robustness test on the watermarked *Harp.wav* signal

		Basic detection	Adaptive synchronization	Frequency alignment	Final watermark
		(BER: %)			w_e
No attack		0	0	—	MARK
SNR (dB)	40 dB	0	0	—	MARK
	36 dB	0	0	—	MARK
	30 dB	0	0	—	MARK
Resampling (22.05 kHz)		×	0	—	MARK
Requantization (8 bit)		0	0	—	MARK
Amplitude	+20 %	0	0	—	MARK
	−20 %	0	0	—	MARK
Lp filtering	8 kHz	0	0	—	MARK
	6 kHz	10.00	0	—	MARK
	5 kHz	15.71	10.00	—	MARK
DA/AD (line-in jack)		0	0	—	MARK
Echo (0.3, 200 ms)		0	0	—	MARK
Reverb (1 s)		0	0	—	MARK
Compression II	96 kbps	17.14	0	—	MARK
	64 kbps	×	0	—	MARK
	48 kbps	×	4.29	—	MARK
Cropping (8 × 25 ms)		×	0	—	MARK
Jittering (0.1 ms/20 ms)		×	0	—	MARK
Inserting (8 × 25 ms)		0	0	—	MARK
PITSM	+4%	2.86	2.86	—	MARK
	+10%	×	×	2.14	MARK
	−4%	×	0	—	MARK
	−10%	×	0.71	—	MARK
TPPSM	+4%	×	6.43	—	MARK
	+10%	×	×	1.43	MARK
	−4%	12.86	4.29	—	MARK
	−10%	×	×	4.29	MARK

Notes: 1. Symbol "×": one detection with a BER of greater than 20 %
2. Symbol "−": one unexecuted advanced detection

requantization, 5 kHz low-pass filtering, and −10 % TPPSM, have a BER of greater than 10 %.

- For different attacks, all the watermarked signals exhibit rather high robustness against most attacks. The improved detection can almost perfectly extract the watermarks from the watermarked signals attacked by amplitude scaling, resampling, DA/AD conversion, echo addition, reverberation, MP3 compression, random samples cropping, jittering, and zeros inserting. It is worth mentioning that our watermarked signals are quite robust against DA/AD conversion, where the BERs of w_e are all equal to 0 %. However, as shown in Chap. 3, most audio watermarking techniques cannot resist DA/AD conversion.

Table 5.7 Results of the basic robustness test on the watermarked *Pop.wav* signal

		Basic detection	Adaptive synchronization	Frequency alignment	Final watermark
		(BER: %)			w_e
No attack		0	0	–	SECURITY
SNR (dB)	40 dB	0.71	0.71	–	SECURITY
	36 dB	3.93	3.93	–	SEСURITY
	30 dB	12.14	10.71	–	SEЧUЯ ITY
Resampling (22.05 kHz)		0	0	–	SECURITY
Requantization (8 bit)		×	19.29	–	SEϾWᴧ1ᴤY
Amplitude	+20 %	0	0	–	SECURITY
	−20 %	0	0	–	SECURITY
Lp filtering	8 kHz	0	0	–	SECURITY
	6 kHz	11.79	2.50	–	SEСURITY
	5 kHz	16.43	9.64	–	SЕϹURIͳY
DA/AD (line-in jack)		13.93	0	–	SECURITY
Echo (0.3, 200 ms)		0	0	–	SECURITY
Reverb (1 s)		13.93	0	–	SECURITY
Compression II	96 kbps	0	0	–	SECURITY
	64 kbps	0	0	–	SECURITY
	48 kbps	2.50	2.50	–	SECURIͳY
Cropping (8 × 25 ms)		×	0	–	SECURITY
Jittering (0.1 ms/20 ms)		×	0	–	SECURITY
Inserting (8 × 25 ms)		0	0	–	SECURITY
PITSM	+4 %	8.57	2.14	–	SECURITY
	+10 %	×	×	9.64	ϿEϹURIͳↃ
	−4 %	×	1.07	–	SECURITY
	−10 %	×	15.36	4.29	SEϹURITY
TPPSM	+4 %	×	10.00	1.43	ϿECURITY
	+10 %	×	×	12.14	ϿEϹURↃͳY
	−4 %	×	14.29	0.36	SECURITY
	−10 %	×	×	8.93	ƷEϹURIͳY

Notes: 1. Symbol "×": one detection with a BER of greater than 20 %
2. Symbol "−": one unexecuted advanced detection

In cases of PITSM and TPPSM attacks, the improved detection can mostly combat PITSM and TPPSM within ±4 %, but fail at larger distortions of ±10 %. Under such circumstances, we resort to the advanced detection with frequency alignment to extract the severely distorted watermarks. From Table 5.4 to Table 5.7, it can be seen that the BERs of w_e attacked by ±10 % PITSM, and TPPSM attacks are reduced greatly after frequency alignment, most of which are not greater than 10 %.

Among various attacks, it is observed that requantization is the most difficult attack. Except for the watermarked *Harp.wav* signal in Table 5.6, the other three watermarked signals are rather vulnerable to the requantization and their BERs are not less than 16 %. Moreover, noise addition poses difficulties for watermark detection as the power of the added noise increases. Note that the decrease of the specified SNR value, i.e., $40\,dB \rightarrow 36\,dB \rightarrow 30\,dB$, indicates the increase of the noise power. To enhance the robustness against these two attacks, higher watermark strengths α_w are required to amplify the magnitude of the watermark signals in the embedding.

In addition to common signal operations and the desynchronization attacks listed above, some combined attacks are also applied on the watermarked signals. The aim is to further evaluate the robustness of the proposed scheme. A combined attack is a combination of two attacks, i.e., $AT_1\,(\cdot)$ followed by $AT_2\,(\cdot)$. The procedure of applying a combined attack on the watermarked signal is as follows: the watermarked signal is first attacked by $AT_1\,(\cdot)$ and the resulting signal is then attacked by $AT_2\,(\cdot)$.

Two types of combined attacks are taken into consideration.

(1) Type I combined attack: $AT_1\,(\cdot)$ is random samples cropping, jittering, or zeros inserting, while $AT_2\,(\cdot)$ is MP3 compression at 96 kbps or low-pass filtering at 8 kHz.
(2) Type II combined attack: $AT_1\,(\cdot)$ is $+5\,\%$ PITSM, $-5\,\%$ PITSM, $+5\,\%$ TPPSM, or $-5\,\%$ TPPSM, while $AT_2\,(\cdot)$ is MP3 compression at 96 kbps, low-pass filtering at 8 kHz, random samples cropping, jittering, or zeros inserting.

Without loss of generality, the watermarked *Bass.wav* signal is used as an example. The coded-image watermark embedded into *Bass.wav* is PROTECTION Table 5.8 shows the results of combined attacks on the watermarked *Bass.wav* signal, including the final extracted watermarks w_e as well as their BERs. Note that all the w_e attacked by Type I combined attacks are extracted by the improved detection. Type II combined attacks are very destructive and the improved detections fail to extract the watermarks. In this case, the advanced detection is employed to recover the w_e attacked by Type II combined attacks.

In Table 5.8, each combined attack is the combination of the attacks on the corresponding row and column. For instance, in (1) Type I combined attacks, the shaded BER (0.86 %) and w_e are the results under the combined attack where $AT_1\,(\cdot)$ is random samples cropping and $AT_2\,(\cdot)$ is MP3 compression at 96 kbps. Also, in (2) Type II combined attacks, the shaded BER (0.57 %) and w_e are the results under the combined attack where $AT_1\,(\cdot)$ is $-5\,\%$ PITSM and $AT_2\,(\cdot)$ is zeros inserting.

From Table 5.8, it can be seen that the proposed scheme is quite resistant to these combined attacks. All the BERs of w_e are less than 10 % and the coded-image watermarks can be clearly identified. It is observed that the combined attacks involving jittering are generally more challenging than the others.

With regard to Type II combined attacks, one point to note is that the length of the PITSM- or TPPSM-attacked signal has been altered by cropping, jittering, and inserting. Even in these cases, the distorted watermarks can be recovered by advanced detection. Therefore, it is proved that a slight change in the length of the attacked signal has no influence on the efficiency of frequency alignment, as discussed in Sect. 4.3.3.2.

Table 5.8 Results of combined attacks on the watermarked *Bass.wav* signal

(1) Type I combined attacks

Cropping, jittering, or inserting followed by compression or filtering

	Compression II (96 kbps)		Lp filtering (8 kHz)	
	BER: %	w_e	BER: %	w_e
Cropping (8×25 ms)	0.86	PROTECTION	1.43	PROTECTION
Jittering (0.1 ms/20 ms)	8.86	PROTECTION	9.14	PROTECTION
Inserting (8×25 ms)	1.14	PROTECTION	1.71	PROTECTION

(2) Type II combined attacks

(a) PITSM followed by compression, filtering, cropping, jittering, or inserting

	PITSM ($+5\%$)		PITSM (-5%)	
	BER: %	w_e	BER: %	w_e
Compression II (96 kbps)	5.71	PROTECTION	4.57	PROTECTION
Lp filtering (8 kHz)	4.29	PROTECTION	8.00	PROTECTION
Cropping (8×25 ms)	1.43	PROTECTION	1.71	PROTECTION
Jittering (0.1 ms/20 ms)	9.43	PROTECTION	6.00	PROTECTION
Inserting (8×25 ms)	1.71	PROTECTION	0.57	PROTECTION

(b) TPPSM followed by compression, filtering, cropping, jittering, or inserting

	TPPSM ($+5\%$)		TPPSM (-5%)	
	BER: %	w_e	BER: %	w_e
Compression II (96 kbps)	5.14	PROTECTION	4.00	PROTECTION
Lp filtering (8 kHz)	7.71	PROTECTION	7.71	PROTECTION
Cropping (8×25 ms)	1.71	PROTECTION	2.29	PROTECTION
Jittering (0.1 ms/20 ms)	7.71	PROTECTION	4.86	PROTECTION
Inserting (8×25 ms)	2.00	PROTECTION	1.14	PROTECTION

5.3.3 Advanced Robustness Test

The advanced robustness test is designed especially for evaluating the proposed
audio watermarking scheme. As described in Sect. 3.1.3.2, the advanced robustness
test is comprised of three parts, namely a test with StirMark for Audio, a test
under collusion, and a test under multiple watermarking. Note that in the advanced
robustness test, all the watermarks are extracted by improved detection, i.e., the
basic detection updated with adaptive synchronization.

5.3.3.1 Test with StirMark for Audio

StirMark for Audio [134] is a publicly available benchmark for robustness evaluation of audio watermarking schemes. In the experiments, we utilize StirMark for
Audio v0.2 with default parameters and a suite of 50 StirMark-attacked signals are
generated accordingly. Note that the attacked signals from StirMark for Audio are

stereo signals. Similar to the method in Sect. 3.1.1, the left channel is taken as the attacked watermarked signal in our scheme.

Based on the description of the attacks in Appendix C and the analysis of the attacked watermarked signals, the following attacks are excluded from the evaluation: *Addfftnoise*, *Extrastereo_30*, *Extrastereo_50*, *Extrastereo_70*, *Nothing*, *Resampling*, and *Voiceremove*. Since the audio test files are monaural, *Extrastereo* attack has no effect. Also, as its name implies, the *Nothing* attack does nothing with the watermarked signals. So in these cases, the watermarked signals remain unchanged and the watermarks can always be extracted perfectly. Moreover, most samples of the attacked signals under *Addfftnoise*, *Resampling*, and *Voiceremove* are zeros, and hence it is unnecessary to proceed with the detection. Other than these, the remaining 43 attacks are included in our experiments. It is worth mentioning that the *Original* attack resembles the original (unattacked) watermarked signal, which is actually the same as the *Nothing* attack. Therefore, the resulting signals from the *Original* attack can be referenced as the original watermarked signals.

Detection results of the watermarked *Bass.wav*, *Gspi.wav*, *Harp.wav*, and *Pop.wav* signals under StirMark for Audio are shown in Table 5.9. For simplicity, the coded images are not illustrated in the table, and we only present the BERs of the extracted watermarks. Apart from the BERs, the ODGs of the attacked signals relative to their host signals are also calculated to get an insight into the amount of distortion caused by the attacks. If the ODG of the attacked signal is comparable to that of the original watermarked signal, it means that the watermarked signal is less affected by this attack. Otherwise, the watermarked signal has already been severely destroyed by the attack.

Table 5.9 shows that the proposed scheme has high resistance to most attacks in StirMark for Audio, including common signal operations and some serious desynchronization attacks, such as *Copysamples*, *Zerolength*, and *Zeroremove*. Although more than half or all four watermark detections fail under the shaded attacks (i.e., *Addnoise_500*, *Addnoise_700*, *Addnoise_900*, *Cutsamples*, and *Zerocross*), these attacks actually have strong negative impact on the fidelity of the watermarked signals. As can be seen, in cases of failed detections, the ODGs are lower than -3.30, while 80 % of the ODGs are even lower than -3.80. Therefore, the attacked signals are very different from the host signals, beyond the premise of the robustness test.

For different watermarked signals, the watermarked *Harp.wav* signal possesses the strongest robustness. Except for one failed detection and the detection under the *Echo* attack, the rest of the attacks cannot destroy the embedded watermarks and the resulting BERs are not more than 5 %. Next comes the watermarked *Pop.wav* signal, which succeeds in 34 detections with BERs of less than 8 % and three detections with BERs of around 16 %. For the watermarked *Bass.wav* signal, the BERs of all 35 surviving watermarks are less than 7 %. Similar to the conclusion in the previous section, the watermarked *Gspi.wav* signal suffers more from the attacks. But even so, the watermarked *Gspi.wav* signal fails in six detections only.

Finally, it should be pointed out that successful detections under *Addbrumm_1100* ~ *Addbrumm_10100* and *Addsinus* attacks are on the condition that the initial

Table 5.9 Results of StirMark for Audio attacks

StirMark attacks	*Bass.wav* BER: %	ODG	*Gspi.wav* BER: %	ODG	*Harp.wav* BER: %	ODG	*Pop.wav* BER: %	ODG
Original	**0**	**−1.986**	**0**	**−1.737**	**0**	**−0.509**	**0**	**−1.076**
Addbrumm_100	0	−2.189	0	−1.891	0	−0.88	0	−1.084
Addbrumm_1100	0	−3.412	0	−3.245	0	−3.109	0	−1.783
Addbrumm_2100	0	−3.635	0	−3.472	0	−3.405	0	−2.610
Addbrumm_3100	0	−3.753	0	−3.580	0	−3.495	0	−3.140
Addbrumm_4100	0	−3.804	0	−3.651	0	−3.565	0	−3.414
Addbrumm_5100	0	−3.830	0	−3.699	0	−3.61	0	−3.560
Addbrumm_6100	0	−3.846	0	−3.732	0	−3.637	0	−3.644
Addbrumm_7100	0	−3.856	0	−3.761	0	−3.658	0	−3.696
Addbrumm_8100	0	−3.863	0	−3.782	0	−3.683	0	−3.730
Addbrumm_9100	0	−3.868	0	−3.798	0	−3.705	0	−3.755
Addbrumm_10100	0	−3.872	0	−3.814	0	−3.712	0	−3.772
Addnoise_100	6.86	−3.763	11.9	−3.870	0	−3.468	3.57	−3.565
Addnoise_300	×	−3.880	11.43	−3.907	0	−3.804	16.07	−3.847
Addnoise_500	×	−3.892	×	−3.911	0	−3.844	×	−3.877
Addnoise_700	×	−3.898	×	−3.911	0	−3.858	×	−3.885
Addnoise_900	×	−3.846	×	−3.912	2.86	−3.863	×	−3.888
Addsinus	0	−3.895	0	−3.819	0	−3.814	0	−3.848
Amplify	0	−2.921	0	−2.613	0	−1.772	0	−2.623
Compressor	0	−3.858	0	−2.058	0	−0.51	0	−1.076
Copysample	×	−3.883	5.71	−3.913	1.43	−3.642	×	−3.624
Cutsamples	×	−3.908	×	−3.913	×	−3.908	×	−3.868
Dynnoise	×	−3.798	×	−3.908	0	−3.14	16.07	−3.855
Echo	×	−3.898	12.86	−3.891	17.14	−3.867	15.36	−3.872
Exchange	1.14	−2.989	5.71	−3.638	3.57	−2.391	7.86	−2.271
FFT_hlpass	0	−3.835	6.19	−3.873	0	−3.373	0	−3.206
FFT_invert	0	−2.128	0	−1.829	0	−0.502	0	−1.085
FFT_real_reverse	0	−2.135	0	−1.934	0	−0.561	0	−1.083
FFT_stat1	4.29	−3.894	7.14	−3.838	2.86	−3.284	2.50	−3.829
FFT_test	4.29	−3.894	7.14	−3.838	2.86	−3.291	2.50	−3.829
Flippsample	0.57	−3.680	5.24	−3.913	0	−2.417	0.36	−3.360
Invert	0	−1.986	0	−1.737	0	−0.509	0	−1.076
Lsbzero	0	−2.011	0	−1.775	0	−0.545	0	−1.097
Normalize	0	−3.317	0	−2.813	0	−3.371	0	−3.357
RC_highpass	0	−2.435	0	−2.092	0	−1.114	0	−1.727
RC_lowpass	0	−2.205	0	−1.759	0	−0.868	0	−1.456
Smooth	1.71	−3.618	19.05	−3.899	0	−2.965	1.43	−2.458
Smooth2	6.29	−3.537	13.81	−3.897	0	−2.884	0	−2.400
Stat1	0	−1.897	0	−1.776	0	−0.58	0	−0.654
Stat2	0	−2.191	0	−1.801	0	−0.826	0	−1.378
Zerocross	×	−3.638	×	−3.896	5.00	−3.171	×	−3.336
Zerolength	0.57	−3.903	0.95	−3.894	0	−2.341	0	−3.867
Zeroremove	0	−3.911	0	−3.909	0	−3.888	0	−3.659

Note: Symbol "×": one detection with a BER of greater than 20 %

watermarking regions are known to the detector. These attacks add high-amplitude buzz or sinus tone throughout the watermarked signal[3], which has an influence on the threshold E_T for the selection of watermarking regions. As a result, the detector cannot properly locate the regions for watermark detection. In this case, E_T must be set at a higher value to select more stable regions. However, as discussed in Sect. 4.1.1, data payload is reduced accordingly.

5.3.3.2 Test Under Collusion

Collusion is one challenging statistical attack on audio watermarking schemes.

Given n watermarked signals $s_w^{(1)}, s_w^{(2)}, \ldots, s_w^{(n)}$ that are generated by separately embedding n watermarks $w_o^{(1)}, w_o^{(2)}, \ldots, w_o^{(n)}$ into host signal s_o, the collusion attack is to create n average watermarked signals $\overline{s_w^{(i)}}$ as follows:

$$
\begin{cases}
s_w^{(j)} = Embedding\left(s_o, w_o^{(j)}\right), & 1 \leq j \leq n \\
\overline{s_w^{(i)}} = \frac{1}{i}\left[s_w^{(1)} + s_w^{(2)} + \ldots + s_w^{(i)}\right], & 1 \leq i \leq n
\end{cases}
\tag{5.9}
$$

In the detection, i watermarks $w_e^{(i,j)}$ are detected from each average watermarked signal $\overline{s_w^{(i)}}$ individually:

$$
w_e^{(i,j)} = Detection\left(\overline{s_w^{(i)}}\right), \quad 1 \leq i \leq n \text{ and } 1 \leq j \leq i
\tag{5.10}
$$

Such an averaging operation weakens the original watermarks and hence makes them hard to detect. Note that the averaging collusion attack is the most common collusion attack and nonlinear collusion attacks [135] are not taken into consideration in the book.

During our experiments, four different watermarks ($w_o^{(1)}$, $w_o^{(2)}$, $w_o^{(3)}$, and $w_o^{(4)}$) are separately embedded into host signal s_o and yield four watermarked signals ($s_w^{(1)}$, $s_w^{(2)}$, $s_w^{(3)}$, and $s_w^{(4)}$). Then, four average watermarked signals are generated, i.e., $\overline{s_w^{(1)}} = s_w^{(1)}$, $\overline{s_w^{(2)}} = \frac{1}{2}\left[s_w^{(1)} + s_w^{(2)}\right]$, $\overline{s_w^{(3)}} = \frac{1}{3}\left[s_w^{(1)} + s_w^{(2)} + s_w^{(3)}\right]$, and $\overline{s_w^{(4)}} = \frac{1}{4}\left[s_w^{(1)} + s_w^{(2)} + s_w^{(3)} + s_w^{(4)}\right]$.

After that, four sets of watermark detections are separately performed as follows:

(1) Detect $w_o^{(1)}$ from $s_w^{(1)}$ to obtain the extracted watermark $w_e^{(1,1)}$.
(2) Detect $w_o^{(1)}$ and $w_o^{(2)}$ separately from $\overline{s_w^{(2)}}$ to obtain the extracted watermarks $w_e^{(2,1)}$ and $w_e^{(2,2)}$.
(3) Detect $w_o^{(1)}$, $w_o^{(2)}$, and $w_o^{(3)}$ separately from $\overline{s_w^{(3)}}$ to obtain the extracted watermarks $w_e^{(3,1)}$, $w_e^{(3,2)}$, and $w_e^{(3,3)}$.

---

[3]In fact, the noises are quite loud already, as proved by the ODGs.

(4) Detect $w_o^{(1)}$, $w_o^{(2)}$, $w_o^{(3)}$, and $w_o^{(4)}$ separately from $s_w^{\overline{(4)}}$ to obtain the extracted watermarks $w_e^{(4,1)}$, $w_e^{(4,2)}$, $w_e^{(4,3)}$, and $w_e^{(4,4)}$.

Table 5.10 shows the results of the averaging collusion attack on *Bass.wav*, *Gspi.wav*, *Harp.wav*, and *Pop.wav*. For *Bass.wav*, the coded-image watermarks used are COLLUSION as $w_o^{(1)}$, WATERMARK as $w_o^{(2)}$, EMBEDDING as $w_o^{(3)}$, and DETECTION as $w_o^{(4)}$. For *Gspi.wav*, the coded-image watermarks used are ROBUST as $w_o^{(1)}$, AUDIO as $w_o^{(2)}$, EMBED as $w_o^{(3)}$, and DETECT as $w_o^{(4)}$. For *Harp.wav*, the coded-image watermarks used are MARK as $w_o^{(1)}$, SIGN as $w_o^{(2)}$, COPY as $w_o^{(3)}$, and HELP as $w_o^{(4)}$. For *Pop.wav*, the coded-image watermarks used are SECURITY as $w_o^{(1)}$, COLLUDE as $w_o^{(2)}$, ATTACKWM as $w_o^{(3)}$, and PROTECT as $w_o^{(4)}$.

To evaluate the perceptual quality of the watermarked signals (including $s_w^{(1)} \sim s_w^{(4)}$ and $s_w^{\overline{(2)}} \sim s_w^{\overline{(4)}}$), their SNRs and ODGs relative to the host signal are calculated. Also, the BERs of the extracted watermarks w_e are calculated to denote the detection rate. Since $w_e^{(1,1)}$ can always be detected without bit errors, the results of $w_e^{(1,1)}$ are omitted in the table.

It is observed from Table 5.10 that for a given host signal, the SNRs and ODGs of $s_w^{\overline{(n)}}$ are generally higher than $s_w^{(n)}$. This means that the average watermarked signals generally have better perceptual quality than the single watermarked signal.

Take *Bass.wav* as an example. On one hand, the SNRs of $s_w^{\overline{(2)}}$, $s_w^{\overline{(3)}}$, and $s_w^{\overline{(4)}}$ are 36.75 dB, 37.54 dB, and 38.65 dB, respectively. These values are higher than the SNRs of $s_w^{(1)}$, $s_w^{(2)}$, $s_w^{(3)}$, and $s_w^{(4)}$, which are 33.45 dB, 33.43 dB, 33.42 dB, and 33.43 dB, respectively. On the other hand, the ODGs of $s_w^{\overline{(2)}}$, $s_w^{\overline{(3)}}$ and $s_w^{\overline{(4)}}$ are -1.064, -0.644, and -0.504, respectively. These values are higher than the ODGs of $s_w^{(1)}$, $s_w^{(2)}$, $s_w^{(3)}$, and $s_w^{(4)}$, which are -1.761, -1.753, -1.829, and -2.001, respectively.

Furthermore, it can seen that the SNRs and ODGs of $s_w^{\overline{(2)}}$, $s_w^{\overline{(3)}}$, and $s_w^{\overline{(4)}}$ increase gradually. This shows that when more copies of the watermarked signals are used in the averaging, the resulting average watermarked signal has a better perceptual quality.

These observations result from the operation of averaging, which weakens the effect of individual watermarks but increases the effect of the host signal. In this way, as more copies are averaged, the average watermarked signal is more similar to the host signal.

Meanwhile, being averaged over more copies, it become more difficult to extract each individual watermark from the average watermarked signals. Remember that $w_e^{(\cdot,1)}$ (including $w_e^{(1,1)}$, $w_e^{(2,1)}$, $w_e^{(3,1)}$, and $w_e^{(4,1)}$) is the distorted $w_o^{(1)}$ extracted respectively from $s_w^{(1)}$, $s_w^{\overline{(2)}}$, $s_w^{\overline{(3)}}$, and $s_w^{\overline{(4)}}$. It can be seen from Table 5.10 that for a given host signal, the BERs of $w_e^{(2,1)}$, $w_e^{(3,1)}$, and $w_e^{(4,1)}$ are usually increasing. For example, the BERs of $w_e^{(1,1)}$, $w_e^{(2,1)}$, $w_e^{(3,1)}$, and $w_e^{(4,1)}$ for *Pop.wav* are 0 %, 0.36 %, 4.64 %, and 6.79 %, respectively.

On the whole, the BERs of all the extracted watermarks are less than 9 %. This indicates that the proposed scheme is quite robust against the averaging collusion attack.

Table 5.10 Results of the averaging collusion attack

		Bass.wav	Gspi.wav	Harp.wav	Pop.wav
(a) Single watermarked signal					
$w_o^{(1)}$		COLLUSION	ROBUST	MARK	SECURITY
$S_w^{(1)}$	SNR/dB	33.45	32.05	24.48	30.41
	ODG	−1.761	−1.757	−0.531	−1.179
$w_o^{(2)}$		WATERMARK	AUDIO	SIGN	COLLUDE
$S_w^{(2)}$	SNR/dB	33.43	32.07	24.45	30.53
	ODG	−1.753	−1.515	−0.574	−1.018
$w_o^{(3)}$		EMBEDDING	EMBED	COPY	ATTACKWM
$S_w^{(3)}$	SNR/dB	33.42	32.03	24.46	30.42
	ODG	−1.829	−1.496	−0.663	−1.079
$w_o^{(4)}$		DETECTION	DETECT	HELP	PROTECT
$S_w^{(4)}$	SNR/dB	33.43	32.04	24.46	30.51
	ODG	−2.001	−1.762	−0.559	−1.084
(b) Average watermarked signal					
$S_w^{\overline{(2)}}$	SNR/dB	36.75	34.90	27.42	33.32
	ODG	−1.064	−1.424	−0.162	−0.453
$S_w^{\overline{(3)}}$	SNR/dB	37.54	36.50	29.10	34.56
	ODG	−0.644	−1.128	−0.087	−0.283
$S_w^{\overline{(4)}}$	SNR/dB	38.65	37.89	30.04	35.31
	ODG	−0.504	−1.046	−0.010	−0.180
$w_e^{(\cdot,1)}$	BER: %	0	2.86	0.71	0.36
	$w_e^{(2,1)}$	COLLUSION	ROBUST	MARK	SECURITY
	BER: %	2.86	4.29	2.86	4.64
	$w_e^{(3,1)}$	COLLUSION	ROBUST	MARK	SECURITY
	BER: %	5.71	8.10	5.00	6.79
	$w_e^{(4,1)}$	COLLUSION	ROBUST	MARK	SECURITY
$w_e^{(\cdot,2)}$	BER: %	0.32	0.57	0	0.41
	$w_e^{(2,2)}$	WATERMARK	AUDIO	SIGN	COLLUDE
	BER: %	2.22	0	0.71	4.90
	$w_e^{(3,2)}$	WATERMARK	AUDIO	SIGN	COLLUDE
	BER: %	6.67	4.57	6.43	8.16
	$w_e^{(4,2)}$	WATERMARK	AUDIO	SIGN	COLLUDE
$w_e^{(\cdot,3)}$	BER: %	4.13	1.71	0.71	3.21
	$w_e^{(3,3)}$	EMBEDDING	EMBED	COPY	ATTACKWM
	BER: %	4.13	2.86	5.00	8.93
	$w_e^{(4,3)}$	EMBEDDING	EMBED	COPY	ATTACKWM
$w_e^{(\cdot,4)}$	BER: %	7.30	5.24	8.57	4.49
	$w_e^{(4,4)}$	DETECTION	DETECT	HELP	PROTECT

5.3.3.3 Test Under Multiple Watermarking

Multiple watermarking is another challenging statistical attack to audio water-marking schemes. However, this attack has received very little attention in most robustness tests.

As described in Sect. 3.1.3.2, multiple watermarking is to sequentially embed n different watermarks $w_o^{(1)}$, $w_o^{(2)}$, ..., $w_o^{(n)}$ in the following way:

$$\begin{cases} s_w^{(1)} = Embedding\left(s_o, w_o^{(1)}\right) \\ s_w^{(i)} = Embedding\left(s_w^{(i-1)}, w_o^{(i)}\right), \quad 2 \leq i \leq n \end{cases} \tag{5.11}$$

In the detection, i watermarks $w_e^{(i,j)}$ are extracted from the watermarked signal $s_w^{(i)}$ individually:

$$w_e^{(i,j)} = Detection\left(s_w^{(i)}\right), \quad 1 \leq i \leq n \ and \ 1 \leq j \leq i \tag{5.12}$$

Similar to the averaging collusion attack, embedding multiple watermarks would also weaken the effect of individual watermarks.

It is worth mentioning that $s_w^{(i)}$ in Eqs. (5.9) and (5.11) differs in the meaning, similar to $w_e^{(i,j)}$ in Eqs. (5.10) and (5.12).

Recall that the same or different audio watermarking techniques can be used in multiple watermarking. For example, the first watermark $w_o^{(1)}$ is embedded using our proposed algorithm. The second watermark can be embedded using the proposed algorithm again or using the echo hiding watermarking. Note that the corresponding detection algorithm must be employed to extract the watermark. Therefore, two types of multiple watermarking experiments are performed, i.e., multiple self-watermarking and inter-watermarking.

- Multiple self-watermarking

In multiple self-watermarking, the host signal s_o is sequentially watermarked n times by one method.

Specifically, we consider that s_o is sequentially watermarked four times ($n = 4$) by the proposed method. The procedure for multiple self-watermarking is described as follows:

(1) Embed $w_o^{(1)}$ into s_o to generate $s_w^{(1)}$ and then detect $w_o^{(1)}$ from $s_w^{(1)}$ to obtain the extracted watermark $w_e^{(1,1)}$.
(2) Embed $w_o^{(2)}$ into $s_w^{(1)}$ to generate $s_w^{(2)}$ and then separately detect $w_o^{(1)}$ and $w_o^{(2)}$ from $s_w^{(2)}$ to obtain the extracted watermarks $w_e^{(2,1)}$ and $w_e^{(2,2)}$.
(3) Embed $w_o^{(3)}$ into $s_w^{(2)}$ to generate $s_w^{(3)}$ and then separately detect $w_o^{(1)}$, $w_o^{(2)}$, and $w_o^{(3)}$ from $s_w^{(3)}$ to obtain the extracted watermarks $w_e^{(3,1)}$, $w_e^{(3,2)}$, and $w_e^{(3,3)}$.
(4) Embed $w_o^{(4)}$ into $s_w^{(3)}$ to generate $s_w^{(4)}$ and then separately detect $w_o^{(1)}$, $w_o^{(2)}$, $w_o^{(3)}$, and $w_o^{(4)}$ from $s_w^{(4)}$ to obtain the extracted watermarks $w_e^{(4,1)}$, $w_e^{(4,2)}$, $w_e^{(4,3)}$, and $w_e^{(4,4)}$.

Table 5.11 shows the results of multiple self-watermarking on *Bass.wav*, *Gspi.wav*, *Harp.wav*, and *Pop.wav*. For *Bass.wav*, the coded-image watermarks used are MULTIPLEWM as $w_o^{(1)}$, ATTACKING as $w_o^{(2)}$, EMBEDDING as $w_o^{(3)}$, and DETECTION as $w_o^{(4)}$. For *Gspi.wav*, the coded-image watermarks used are MULTI as $w_o^{(1)}$, ATTACK as

Table 5.11 Results of multiple self-watermarking

		Bass.wav	Gspi.wav	Harp.wav	Pop.wav
$w_o^{(1)}$		MULTIPLEWM	MULTI	MARK	SECURITY
$S_w^{(1)}$	SNR/dB	33.36	32.04	24.47	30.46
	ODG	-2.027	-1.516	-0.475	-1.104
$w_o^{(2)}$		ATTACKING	ATTACK	SIGN	MULTIPLE
$S_w^{(2)}$	SNR/dB	29.79	28.49	20.85	26.66
	ODG	-3.131	-2.672	-1.702	-1.959
$w_o^{(3)}$		EMBEDDING	EMBED	COPY	ATTACKWM
$S_w^{(3)}$	SNR/dB	27.93	26.55	18.05	23.79
	ODG	-3.391	-3.291	-2.808	-2.756
$w_o^{(4)}$		DETECTION	DETECT	HELP	PROTECT
$S_w^{(4)}$	SNR/dB	26.24	25.18	16.08	21.86
	ODG	-3.591	-3.564	-3.339	-3.240
$w_e^{(\cdot,1)}$	BER: %	0	0	0	0
	$w_e^{(1,1)}$	MULTIPLEWM	MULTI	MARK	SECURITY
	BER: %	0.29	2.86	0	0.36
	$w_e^{(2,1)}$	MULTIPLEWM	MULTI	MARK	SECURITY
	BER: %	1.43	7.43	2.86	1.79
	$w_e^{(3,1)}$	MULTIPLEWM	MULTI	MARK	SECURITY
	BER: %	4.00	7.43	2.86	2.14
	$w_e^{(4,1)}$	MULTIPLEWM	MULTI	MARK	SECURITY
$w_e^{(\cdot,2)}$	BER: %	0	0	0	0
	$w_e^{(2,2)}$	ATTACKING	ATTACK	SIGN	MULTIPLE
	BER: %	0.32	0	0	1.07
	$w_e^{(3,2)}$	ATTACKING	ATTACK	SIGN	MULTIPLE
	BER: %	3.17	9.52	0	1.43
	$w_e^{(4,2)}$	ATTACKING	ATTACK	SIGN	MULTIPLE
$w_e^{(\cdot,3)}$	BER: %	0	0	0	0
	$w_e^{(3,3)}$	EMBEDDING	EMBED	COPY	ATTACKWM
	BER: %	0	1.14	0	2.14
	$w_e^{(4,3)}$	EMBEDDING	EMBED	COPY	ATTACKWM
$w_e^{(\cdot,4)}$	BER: %	0	0	0	0
	$w_e^{(4,4)}$	DETECTION	DETECT	HELP	PROTECT

$w_o^{(2)}$, EMBED as $w_o^{(3)}$, and DETECT as $w_o^{(4)}$. For *Harp.wav*, the coded-image watermarks used are MARK as $w_o^{(1)}$, SIGN as $w_o^{(2)}$, COPY as $w_o^{(3)}$, and HELP as $w_o^{(4)}$. For *Pop.wav*, the coded-image watermarks used are SECURITY as $w_o^{(1)}$, MULTIPLE as $w_o^{(2)}$, ATTACKWM as $w_o^{(3)}$, and PROTECT as $w_o^{(4)}$.

For evaluation purposes, we calculate the SNRs and ODGs of the watermarked signals (including $s_w^{(1)} \sim s_w^{(4)}$) relative to the host signal, as well as the BERs of the extracted watermarks.

It is observed from Table 5.11 that for a given host signal, the SNRs and ODGs of $s_w^{(1)}$, $s_w^{(2)}$, $s_w^{(3)}$, and $s_w^{(4)}$ decrease gradually. This means that the perceptual quality gets worse if the signal is watermarked more times. Take *Bass.wav* as an example. The decreasing SNRs of $s_w^{(1)}$, $s_w^{(2)}$, $s_w^{(3)}$, and $s_w^{(4)}$ are 33.36 dB, 29.79 dB, 27.93 dB, and 26.24 dB, respectively. Also, the decreasing ODGs of $s_w^{(1)}$, $s_w^{(2)}$, $s_w^{(3)}$, and $s_w^{(4)}$ are $-2.027, -3.131, -3.391$, and -3.591, respectively. The reason is that more samples of the signal are modified during the embedding of more watermarks.

Meanwhile, because of being watermarked more times, each individual watermark becomes more difficult to extract from the multiple watermarked signals. Remember that $w_e^{(\cdot,1)}$ (including $w_e^{(1,1)}$, $w_e^{(2,1)}$, $w_e^{(3,1)}$, and $w_e^{(4,1)}$) is the distorted $w_o^{(1)}$ extracted respectively from $s_w^{(1)}$, $s_w^{(2)}$, $s_w^{(3)}$, and $s_w^{(4)}$. It can be seen from Table 5.11 that for a given host signal, the BERs of $w_e^{(2,1)}$, $w_e^{(3,1)}$, and $w_e^{(4,1)}$ are usually increasing. For example, the BERs of $w_e^{(1,1)}$, $w_e^{(2,1)}$, $w_e^{(3,1)}$, and $w_e^{(4,1)}$ for *Pop.wav* are 0 %, 0.36 %, 1.79 %, and 2.14 %, respectively.

On the whole, the BERs of all the extracted watermarks are less than 10 %. This indicates that the proposed scheme is quite robust against multiple self-watermarking.

- Inter-watermarking

In inter-watermarking, the watermarked signal is separately re-watermarked by other audio watermarking techniques.

In addition to the proposed method ("Proposed"), four watermarking techniques in Chap. 3 are also considered, i.e., cepstrum domain watermarking ("Cepstrum"), wavelet domain watermarking ("Wavelet"), echo hiding with kernel 3 ("Echo"), and histogram-based watermarking ("Histogram"). Note that least significant bit (LSB) modification, phase coding, and spread spectrum (SS) watermarking are not taken into consideration, since these three watermarking techniques cannot preserve the perceptual quality of the watermarked signals.

During the process of inter-watermarking, cepstrum domain watermarking, wavelet domain watermarking, echo hiding, and histogram-based watermarking use the parameters as specified in Figs. 3.8, 3.10, 3.13, and 3.14, respectively. The results in Chap. 3 show that these parameters provide the best performance for each watermarking technique. Also, instead of the coded-image watermarks, the PRSs are directly embedded at full capacity. Moreover, to avoid perceived noise caused by watermarking the silence, the PRSs are embedded into the watermarking regions selected by the proposed method.

To make a comparison of robustness against inter-watermarking between the proposed method and the other audio watermarking techniques, two experiments are carried out. The first experiment is to evaluate the ability of the proposed method resistance to inter-watermarking. The second experiment is to evaluate the ability of the other watermarking techniques resistance to inter-watermarking.

✳ *In Experiment I, the watermarked signal generated by the proposed method is separately re-watermarked by the considered watermarking techniques.*

The procedure for Experiment I is described as follows.

(1) The proposed method embeds w_o into s_o to generate s_w and then detects w_o from s_w to obtain the extracted watermark w_e.

(2) s_w is re-watermarked by the proposed method (i.e., self-watermarking) and s_w^{prop} is generated. Then, the proposed method detects w_o from s_w^{prop} to obtain the extracted watermark w_e^{prop}.

(3) s_w is re-watermarked by cepstrum domain watermarking and s_w^{ceps} is generated. Then, the proposed method detects w_o from s_w^{ceps} to obtain the extracted watermark w_e^{ceps}.

(4) s_w is re-watermarked by wavelet domain watermarking and s_w^{wave} is generated. Then, the proposed method detects w_o from s_w^{wave} to obtain the extracted watermark w_e^{wave}.

(5) s_w is re-watermarked by echo hiding and s_w^{echo} is generated. Then, the proposed method detects w_o from s_w^{echo} to obtain the extracted watermark w_e^{echo}.

(6) s_w is re-watermarked by histogram-based watermarking and s_w^{hist} is generated. Then, the proposed method detects w_o from s_w^{hist} to obtain the extracted watermark w_e^{hist}.

Table 5.12 shows the results of inter-watermarking Experiment I on *Bass.wav*, *Gspi.wav*, *Harp.wav*, and *Pop.wav* signals. The coded-image watermarks embedded are MULTIPLEWM for *Bass.wav*, MULTI for *Gspi.wav*, MARK for *Harp.wav*, and SECURITY for *Pop.wav*.

For evaluation purposes, we calculate the SNRs and ODGs of the watermarked signals (including s_w and $s_w^{prop} \sim s_w^{hist}$) relative to the host signal, as well as the BERs of the extracted watermarks.

Table 5.12 shows that most watermarks after being inter-watermarked can be perfectly extracted. Only when the watermarked *Bass.wav*, *Gspi.wav*, and *Pop.wav* signals are re-watermarked by the proposed method or histogram-based watermarking, the extracted watermarks are slightly distorted. Yet the BERs are still less than 3 %.

Moreover, for a given host signal, the ODGs of s_w^{ceps} and s_w^{wave} are comparable to that of s_w. However, the ODGs of s_w^{prop}, s_w^{echo}, and s_w^{hist} are usually much lower. Take *Pop.wav* as an example. The ODGs of s_w^{ceps} and s_w^{wave} are -1.568 and -1.161, not much lower than the ODG of s_w, being -1.104. However, the ODGs of s_w^{prop}, s_w^{echo}, and s_w^{hist} are quite different, being -1.959, -2.190, and -2.305 respectively. This indicates that cepstrum domain watermarking and wavelet domain watermarking have less influence on the perceptual quality than other techniques, which agrees with the results in Sects. 3.2.4 and 3.2.5.

It is also found that for a given host signal, the SNRs are not in accordance with the ODGs. Take *Pop.wav* as an example. The ODGs of s_w and s_w^{wave} are similar, i.e., -1.104 and -1.161. However, their SNRs are quite different, i.e., 30.46 dB and 21.35 dB. As already mentioned in Sects. 3.2.5 and 5.2.2, such an observation motivates us to investigate other objective quality measures in Chap. 6.

✳ *In Experiment II, the watermarked signals generated by the considered watermarking techniques are separately re-watermarked by these techniques.*

The above five watermarking techniques are employed as *host techniques* as well as *attack techniques*, i.e., the proposed method ("Proposed"), cepstrum domain watermarking ("Cepstrum"), wavelet domain watermarking ("Wavelet"),

Table 5.12 Results of inter-watermarking Experiment I

		Bass.wav	Gspi.wav	Harp.wav	Pop.wav
w_o		MULTIPLEWM	MULTI	MARK	SECURITY
s_w	SNR/dB	33.36	32.04	24.47	30.46
	ODG	−2.027	−1.516	−0.475	−1.104
	BER: %	0	0	0	0
	w_e MULTIPLEWM	MULTI	MARK	SECURITY	
Proposed s_w^{prop}	SNR/dB	29.79	28.49	20.85	26.66
	ODG	−3.131	−2.672	−1.702	−1.959
	BER: %	0.29	2.86	0	0.36
	w_e^{prop}	MULTIPLEWM	MULTI	MARK	SECURITY
Cepstrum s_w^{ceps}	SNR/dB	20.12	20.13	13.68	15.26
	ODG	−2.400	−1.651	−0.583	−1.568
	BER: %	0	0	0	0
	w_e^{ceps}	MULTIPLEWM	MULTI	MARK	SECURITY
Wavelet s_w^{wave}	SNR/dB	25.23	24.22	18.25	21.35
	ODG	−2.262	−1.774	−0.541	−1.161
	BER: %	0	0	0	0
	w_e^{wave}	MULTIPLEWM	MULTI	MARK	SECURITY
Echo s_w^{echo}	SNR/dB	11.07	10.26	10.79	12.09
	ODG	−2.194	−2.686	−2.157	−2.190
	BER: %	0	0	0	0
	w_e^{echo}	MULTIPLEWM	MULTI	MARK	SECURITY
Histogram s_w^{hist}	SNR/dB	33.06	31.85	24.42	30.00
	ODG	−3.613	−2.970	−1.465	−2.305
	BER: %	1.14	1.14	0	0
	w_e^{hist}	MULTIPLEWM	MULTI	MARK	SECURITY

Note: Symbol "⇑": the column identical to the one in Table 5.13 below

echo hiding ("Echo"), and histogram-based watermarking ("Histogram"). Each host technique embeds the watermark into the host signal to generate the watermarked signal s_w. Then, s_w is separately re-watermarked by the attack techniques and we get the re-watermarked signals accordingly. Later, the host technique detects the watermark from each re-watermarked signal separately.

Clearly, inter-watermarking Experiment I is one special case of inter-watermarking Experiment II, in which the proposed method is the host technique.

Without loss of generality, *Bass.wav* is taken as an example. Table 5.13 shows the results of inter-watermarking Experiment II on *Bass.wav*. Specifically, the shaded cases on the diagonal line refer to each technique's self-watermarking. For evaluation purposes, we calculate the SNRs and ODGs of the watermarked signals relative to the host signal, as well as the BERs of the extracted watermarks.

Since inter-watermarking Experiment I is one special case of inter-watermarking Experiment II, the column indicated by "⇑" in Tables 5.13 and 5.12 shares the same results.

Table 5.13 Results of inter-watermarking Experiment II on *Bass.wav* signal

			Host technique				
			Proposed	Cepstrum	Wavelet	Echo	Histogram
	Watermark length N_w		350	324	324	162	40
	s_w	SNR/dB	33.36	20.63	26.01	10.88	44.56
		ODG	−2.027	−0.602	−0.557	−2.200	−2.123
		BER: %	0	0	0	0	0
Attack	Proposed	SNR/dB	29.79	20.30	25.21	10.84	32.99
technique		ODG	−3.131	−2.162	−2.260	−2.190	−3.598
		BER: %	0.29	0	0	0	0
	Cepstrum	SNR/dB	20.12	19.61	19.94	10.48	20.61
		ODG	−2.400	−0.527	−0.534	−2.200	−2.141
		BER: %	0	×	×	0	×
	Wavelet	SNR/dB	25.23	24.60	24.57	10.78	25.85
		ODG	−2.262	−0.574	−0.604	−2.200	−2.149
		BER: %	0	×	×	0	×
	Echo	SNR/dB	11.07	10.55	10.99	7.85	11.19
		ODG	−2.194	−2.200	−2.200	−2.190	−2.200
		BER: %	0	0	0	0.62	×
	Histogram	SNR/dB	33.06	20.62	25.93	10.87	40.05
		ODG	−3.613	−2.031	−2.095	−2.200	−2.938
		BER: %	1.14	0	0	0.62	×

Notes: 1. Symbol "×": one detection with a BER of greater than 20 %
2. Symbol "⇑": the column identical to the one in Table 5.12 above

From Table 5.13, it is observed that only the proposed method and echo hiding are robust against inter-watermarking by all five watermarking techniques. The BERs of the extracted watermarks are less than 2 %. Note that the successful detection of echo hiding's self-watermarking is conditional: the echo delays used by the attack technique are different (as far away as possible) from the ones used by the host technique.

Meanwhile, the other three techniques fail in some cases of inter-watermarking. For example, given the watermarked signal generated by cepstrum domain watermarking, the embedded watermark cannot survive the re-watermarking by its self-watermarking or wavelet domain watermarking. Similarly, given the watermarked signal generated by wavelet domain watermarking, the embedded watermark cannot survive the re-watermarking by its self-watermarking or cepstrum domain watermarking. By contrast, histogram-based watermarking shows the weakest resistance to inter-watermarking. Given the watermarked signal generated by histogram-based watermarking, the embedded watermark can merely survive the re-watermarking by the proposed method.

In summary, the proposed audio watermarking scheme performs well throughout the robustness test.

5.4   Security Analysis

The goal of security analysis is to evaluate the security level of the proposed audio watermarking scheme. As discussed in Sect. 3.2, cepstrum domain watermarking and wavelet domain watermarking that are based on statistical mean manipulation (SMM), echo hiding, and histogram-based watermarking all suffer from security problem to varying degrees. A theoretical analysis of watermarking security is not the focus of this book. As introduced in Sect. 1.3.2.3, an intuitive method of security analysis is to calculate the possible ways for embedding. If there were more possible ways for embedding, unauthorized detection without secret keys would become more difficult to identify and/or remove the embedded watermark.

In our experiments, each block is divided into 32 nonlinear subbands, where 28 subbands are randomly selected for embedding. In this case, $N_{subband} = 32$ and $\tilde{N}_{subband} = 28$. Accordingly, the number of possible ways due to channel scrambling is calculated by using Eq. (4.8):

$$
\begin{aligned}
N_{scrambling} &= P\left(N_{subband},\ \tilde{N}_{subband}\right) = \frac{N_{subband}!}{\left(N_{subband}-\tilde{N}_{subband}\right)!} \\
&= P(32,\ 28) = \frac{32!}{(32-28)!} \approx 1.1 \times 10^{34}
\end{aligned}
\tag{5.13}
$$

Such a huge number (i.e., 1.1×10^{34}) makes unauthorized detection nearly impossible, which means that the property of the security has increased greatly. This is just one code complexity, which can be further multiplied by the complexity introduced by the PRNs.

5.5   Data Payload and Computational Complexity

Data payload and computational complexity are two criteria of minor consideration in audio watermarking for copyrights protection.

5.5.1   Estimation of Data Payload

As defined in Sect. 1.3.1.4, data payload (or capacity) of one audio watermarking scheme is the number of bits embedded into a one-second audio fraction. According to the embedding algorithm, the data payload of the proposed audio watermarking scheme, DP_B, is expressed as follows:

$$
DP_B = \frac{2f_s \cdot N_{bit}}{N \cdot N_c \cdot N_{unit}}\ bps
\tag{5.14}
$$

where f_s is the sampling frequency of audio signal, N_{bit} is the number of watermark bits embedded per block, N is the frame length, N_c is the number of frames per unit, and N_{unit} is the number of units per block. Note that factor 2 in Eq. (5.14) is due to half-overlapping between adjacent audio frames.

Furthermore, if the coded-image watermark is adopted, the data payload in terms of letters, DP_L, is

$$DP_L = \frac{DP_B}{L_w}\, lps \tag{5.15}$$

where $L_w = 35$ is the number of bits comprising one letter and DP_L is expressed in letter per second (lps).

Note that the data payload discussed above refers to the theoretical data payload of one audio watermarking scheme, which solely depends on the watermark embedder. That is, once the embedding parameters and the embedding algorithm used by the watermark embedder are chosen, theoretical data payload is determined subsequently.

The values of these experiment parameters determined in Sect. 5.1 are $N = 512$, $N_c = 4$, $N_{unit} = 10$, and $N_{bit} = 4$. Moreover, all the audio test files are sampled at 44.1 kHz, i.e., $f_s = 44.1$ kHz. Therefore, the data payload of the scheme under evaluation is equal to

$$DP_B = \frac{2 * 44100 * 4}{512 * 4 * 10} \approx 17.2\, bps \tag{5.16}$$

$$DP_L = \frac{17.2}{35} \approx 0.5\, lps \tag{5.17}$$

which are sufficient for the purpose of copyrights protection.

From Table 5.14, however, it is observed that the watermarks embedded in different host signals have quite different lengths, although the same watermark embedder is employed. Thus, the practical data payload $(\widetilde{DP}_B)$ is defined as the watermark length divided by the duration of the audio signal. For example, we calculate the practical data payload for *Bass.wav*, 350 bits/24.9 s = 14.1 bps; *Gspi.wav*, 210 bits/19 s = 11.1 bps; *Harp.wav*, 140 bits/16.4 s = 8.5 bps; and *Pop.wav*, 280 bits/20 s = 14 bps. The duration of all the test signals is listed in Appendix D. By averaging these four values, the average practical data payload of the proposed scheme is considered to be 11.9 bps.

The practical data payload of one host signal depends on the watermark embedder as well as the selected watermarking regions of each host signal. If a host signal contains more silences and trifle fractions, the watermarking regions are of smaller size and hence the practical data payload is lower. Obviously, the practical data payload cannot always exceed the theoretical data payload. As the practical data payload is approaching the theoretical value, more samples of the host signal are used for watermarking.

Table 5.14 Results of the computational complexity estimation

	Bass.wav	Gspi.wav	Harp.wav	Pop.wav
Duration of host signal (sec)	24.9	19.0	16.4	20.0
Watermark length (bits)	350	210	140	280
Embedding time (sec)	256.1	158.7	97.0	196.3
Average embedding speed (bps)	1.37	1.32	1.44	1.43
Detection time (sec)	56.7	17.3	11.9	31.2
Average detection speed (bps)	6.17	12.14	8.15	8.97

5.5.2   *Estimation of Computational Complexity*

As mentioned in Sect. 1.3.1.5, computational complexity is evaluated in terms of the speed, which is further denoted as the embedding time $\left(t_{embedding}\right)$ and detection time $(t_{detection})$ relative to the duration of the host audio signal. Moreover, average embedding and detection speeds are also employed to indicate the rate of embedding and detection. That is, if a total of N_w watermark bits are embedded in time $t_{embedding}$, the average embedding speed is given by

$$CC_{embedding} = \frac{N_w}{t_{embedding}} \, bps \tag{5.18}$$

Similarly, if N_w watermark bits are detected in time $t_{detection}$, the average detection speed is given by

$$CC_{detection} = \frac{N_w}{t_{detection}} \, bps \tag{5.19}$$

Note that the average embedding and detection speeds are expressed in the same unit as data payload, i.e., bps.

For the test platform, all the experiments are conducted on a Pentium 4 2.4 GHz computer with 1 GB RAM. Table 5.14 shows the results of the computational complexity estimation on *Bass.wav*, *Gspi.wav*, *Harp.wav*, and *Pop.wav*. Note that for the detections measured in the experiments, watermark bits are detected from the watermarked signals without being attacked.

From Table 5.14, it is observed that although different host signals differ with embedding time, their average embedding speeds are similar. On the other hand, the average detection speed of different watermarked signals varies in the detection time as well as the average detection speed. For example, the average detection speed of the watermarked *Gspi.wav* signal is almost twice as fast as than that of the watermarked *Bass.wav* signal. This is due to different implementation mechanisms of the embedding and detection algorithms in MATLAB.

During the execution of the embedding algorithm, the host signal is processed block by block to embed watermark bits. According to Fig. 4.3, block size is merely determined by N, N_c, and N_{unit}, not related to the host signal. Therefore, different host signals have the same utilization of computer memory in the embedding, resulting in a similar average embedding speed.

However, during the execution of the detection algorithm, the watermark bits are extracted from each watermarking region. The reason for this is that in the cases of desynchronization attacks, the conformation of blocks is distorted. Therefore, the magnitudes of all tiles in every watermarking region (not just in every block) need to be provided simultaneously for block synchronization, as shown in Sect. 4.3.1. Consequently, different host signals have a different utilization of computer memory in the detection. Moreover, large watermarking regions demand more computer memory, thus the average detection speed is slow and vice versa for small watermarking regions.

On the whole, the average detection speed is much faster than the average embedding speed, which is a desirable attribute in copyrights protection application.

5.6 Performance Comparison

Table 5.15 compares the performance of our proposed scheme ("Proposed") with several existing audio watermarking schemes, sorted by chronological order. The chosen schemes were not implemented in the book. Therefore, if the result is not reported in the publication, it is marked by symbol / in the table. Also if the published result is obtained or interpreted in a different way, it is marked by the symbol $*$.

The investigation is focused on imperceptibility ("*Impcpty*"), robustness, and data payload ("*Payload*"), since security and computational complexity were not taken into consideration by most schemes.

- For imperceptibility evaluation, the commonly used SNR is employed as the metric. Moreover, it usually refers to the average SNR of all the watermarked signals, since there are a number of host audio signals adopted in every scheme.
- For the robustness test, the attacks include noise addition ("NA"), resampling ("RS"), amplitude scaling ("AM"), low-pass filtering ("LP"), echo addition ("ECHO"), MP3 compression ("MP3"), and PITSM ("TSM").[4] Other attacks such as requantization, DA/AD conversion, reverberation, random samples cropping, jittering, zeros inserting, and TPPSM are not listed in the table, because they were either performed in a varying way or not even conducted in most

---

[4]The attacks with symbol $*$ in Table 5.15 are described as follows. Under the "NA" category, the schemes in [5, 7] did not specify the value of the SNR. Under the "AM" category, the schemes in [5, 7] compressed the amplitude with a nonlinear gain function. Under the "LP" category, the schemes in [3, 8] tested band-pass filtering only. Under the "TSM" category, the schemes in [9, 10] implemented random stretching (at $\pm 4\,\%$ and $\pm 8\,\%$, respectively) merely by omitting or inserting a random number of samples, which is considered similar to random sample cropping/inserting.

Table 5.15 Performance comparison of different audio watermarking schemes

	Impcpty	Robustness							Payload
	(SNR)	NA	RS	AM	LP	ECHO	MP3	TSM	$(\widetilde{DP}_B)$
Proposed	30.1 dB	30 dB	22.05 kHz	±20 %	5 kHz	(0.3, 200 ms)	48 kbps	±10 %	11.9 bps
[1]	33.5 dB	5 dB	8.82 kHz	/	5 kHz	/	32 kbps	±1 %	5.4 bps
[2]	43.8 dB	40 dB	16 kHz	±20 %	7 kHz	/	64 kbps	±25 %	2 bps
[3]	*	20 dB	24 kHz	+50 %	*	(0.4, 100 ms)	64 kbps	/	<10 bps
[4]	/	20 dB	22.05 kHz	*	5 kHz	(0.5, 100 ms)	32 kbps	±5 %	8.4 bps
[5]	29.5 dB	*	22.05 kHz	+50 %	4 kHz	(0.4, 100 ms)	32 kbps	±10 %	4.3 bps
[6]	/	36 dB	48 kHz	/	4 kHz	/	56 kbps	±10 %	2.3 bps
[7]	32.4 dB	*	22.05 kHz	*	8 kHz	(0.5, 100 ms)	48 kbps	±3 %	11.8 bps
[8]	/	36 dB	/	/	*	(/, 100 ms)	32 kbps	±4 %	0.5-1 bps
[9, 10]	35 dB	40 dB	/	/	/	(0.5, 100 ms)	96 kbps	*	2.2 bps

schemes.[5] It is worth mentioning that we do not compare the detailed results of the BER between the schemes, but list their tolerance for each attack that was reported in their publications. The reason is that different host audio signals were used in different schemes, and moreover, some schemes calculated the BERs in a different way. Thus, there is no direct comparison in terms of the BER.

- For data payload estimation, the practical data payload $(\widetilde{DP}_B)$ instead of the theoretical data payload is adopted for comparison. Since most schemes have no theoretical analysis on the data payload, the actual amount of watermark bits embedded into one host signal of certain duration is calculated as the practical data payload. Similar to the SNRs, the practical data payloads shown in the table are usually the average values of all the watermarked signals in each scheme.

Table 5.15 shows that these schemes have different performance characteristics. On the average, the proposed scheme achieves the best compromise between imperceptibility, robustness, and capacity.

- In terms of imperceptibility, the SNR of the proposed scheme is within the range of other schemes. Also, the average ODG of the proposed scheme is -1.33, obtained from Table 5.2. In spite of a higher SNR, the ODGs reported in the scheme in [2] are around -1.80, not superior to the proposed scheme. Without addressing the SNR, the scheme in [3] has an average ODG of -0.93.
- In terms of robustness, the focus is on the performance under PITSM, as well as noise addition, low-pass filtering, and MP3 compression, since most schemes show high resistance to resampling, amplitude scaling, and echo addition. Under PITSM, only the proposed scheme and the schemes in [2, 5, 6] can resist excessive distortion of up to $\pm 10\%$ or greater, and hence are chosen for further comparison. The proposed scheme is robust against PITSM ($\pm 10\%$), noise addition (30 dB), low-pass filtering (5 kHz), and MP3 compression (48 kbps). Compared to the proposed scheme, the scheme in [2] is quite robust against PITSM ($\pm 25\%$), but relatively vulnerable to noise addition (40 dB), low-pass filtering (7 kHz), and MP3 compression (64 kbps). The scheme in [5] is slightly more robust against Low-pass filtering (4 kHz) and MP3 compression (32 kbps); nevertheless the SNR has no specified value to compare the robustness against noise addition.[6] The scheme in [6] is slightly more robust against low-pass

---

[5]These unlisted attacks were undertaken in several schemes as follows. Requantization: only the scheme in [3] tested 8-bit requantization and the detection succeeded. DA/AD conversion: the schemes in [4, 7–10] tested DA/DA conversion and the detections succeeded. Cropping: the schemes in [2, 4, 5] tested different cropping operations and the detections succeeded. Jittering: the schemes in [2, 5] tested different jittering operations and the detections succeeded. TPPSM: the scheme in [1] tested $\pm 1\%$ pitch-scaling and the detection succeeded; the scheme in [3] tested the case that the pitch is shifted up by two semitones and the detections completely failed; the schemes in [9, 10] implemented pitch shifting (at $\pm 4\%$ and $\pm 8\%$ respectively) merely by linear interpolation without anti-alias filtering and the detections succeeded.

[6]It was reported as "noise addition that can be heard clearly by everybody [5]."

filtering (4 kHz), but less against noise addition (36 dB) and MP3 compression (56 kbps).
- In terms of the data payload, the proposed scheme has the highest practical data payload among these schemes, i.e., 11.9 bps. In particular, this value is much higher than data payloads of the schemes in [2, 5, 6], i.e., 2 bps, 4.3 bps, and 2.3 bps respectively. Moreover, as shown in Eq. (5.16), the theoretical data payload of the proposed scheme is even higher—about 17.2 bps.

5.7 Summary

In this chapter, the performance of the proposed audio watermarking scheme has been thoroughly evaluated with respect to imperceptibility, robustness, security, data payload, and computational complexity. Specifically, the designed performance evaluation consists of perceptual quality assessment, robustness test, security analysis, estimations of the data payload, and computational complexity. Without loss of generality, the performance evaluation presented in this chapter can serve as one comprehensive benchmark of audio watermarking algorithms.

Firstly, the subjective listening test and the objective evaluation test were employed in the perceptual quality assessment. Specifically, the subjective listening test includes the MUSHRA test and SDG rating, while the objective evaluation test includes the calculation of the ODG (using PEAQ) and the SNR value. Secondly, both basic and advanced robustness tests were carried out. Basic robustness test includes common signal operations (e.g., noise addition, resampling, requantization, amplitude scaling, low-pass filtering, DA/AD conversion, echo addition, reverberation, and MP3 compression), desynchronization attacks (e.g., random samples cropping, jittering, zeros inserting, PITSM, and TPPSM), and combined attacks (e.g., Type I and Type II combined attacks). The advanced robustness test includes StirMark for Audio, averaging collusion, and multiple watermarking (e.g., self-watermarking and two types of inter-watermarking). Thirdly, the number of possible embedding ways due to channel scrambling was calculated in the security analysis. Furthermore, both theoretical and practical data payloads were calculated. Finally, computational complexity was evaluated in terms of the embedding/detection PC computing time as well as the average embedding/detection speed.

The experimental results show that the watermarked audio signals are perceptually transparent, robust against various attacks, and self-secured from unauthorized detection. Also, watermarking efficiency of the proposed technique is satisfactory with respect to the data payload and computational complexity as compared to the other methods.

Compared with other reported schemes, the proposed scheme achieves a better compromise between imperceptibility, robustness, and data payload. Thus, it is concluded that the proposed audio watermarking scheme performs well for the purpose of copyrights protection.

Chapter 6
Perceptual Evaluation Using Objective Quality Measures

Imperceptibility is a prerequisite to the use of the watermarked audio; hence, perceptual quality assessment is worthy of more attention. Objective quality measures have been widely used in speech quality evaluation. In this chapter, we introduce objective quality measures used for the first time in the perceptual quality evaluation of audio watermarking.

Perceptual quality assessment in audio watermarking including subjective listening tests and objective evaluation tests is reviewed first. This is followed by a description of the objective quality measures under investigation. Next, several experiments are performed to explore the relations between objective quality measures and perceptual quality in the context of different audio watermarking techniques. Finally, some comments are made to summarize the performance of the considered objective quality measures as the perceptual quality predictors.

6.1 Perceptual Quality Evaluation

The aim of audio watermarking is to embed an imperceptible, robust, and secure watermark into host signals. From the viewpoint of communication theory, the watermark is inserted into a cover signal like a kind of noise. Considering that the process of watermarking should be perceptually transparent, the perceptual quality of the watermarked audio signal is evaluated relative to the host audio signal.

As described in Chap. 1, there are two approaches to perceptual quality assessment of audio watermarking: (1) subjective listening tests by human acoustic perception and (2) objective evaluation tests by perception modelling or quality measures.

- Subjective listening test

In the subjective listening tests, the listeners are asked to compare the perceptual quality of the watermarked audio signal with the host audio signal. As stated in

Y. Lin and W.H. Abdulla, *Audio Watermark: A Comprehensive Foundation Using MATLAB*, 159
DOI 10.1007/978-3-319-07974-5_6, © Springer International Publishing Switzerland 2015

Sect. 1.3.2.1, the ABX test and the MUSHRA test (i.e., MUlti Stimuli with Hidden Reference and Anchors) are two commonly used methods.

In the ABX listening test (see Appendix B), the listener has to identify an unknown sample X as being A or B, with A (the host signal) and B (the watermarked signal) available for reference. Initially, the ABX test was designed for the assessment of small deterioration [43]. Note that ABX tests can also be performed as ABC/HR tests, i.e., double blind, triple stimulus, with hidden reference [31]. Specifically, stimulus A is the host signal for reference, whereas stimulus B and C are the host and watermarked signals in randomized order. After listening to three stimuli, the listener is asked to decide between B and C as the hidden reference signal, and then the remaining one is the watermarked signal. Finally, the watermarked signal is evaluated relative to the host signal by using a subjective difference grade (SDG), as described in Table 1.2.

The MUSHRA test (see Sect. 5.2.1) is developed for assessing intermediate audio quality [44]. Since multiple stimuli including the hidden reference and a few additional signals (anchors) are employed, the MUSHRA test is supposed to be more reliable than the ABX test in the presence of slightly larger distortions.

Subjective listening tests are indispensable to perceptual quality assessment, since the ultimate judgment is made by human perception. However, it is quite time-consuming and cost intensive to conduct such listening tests. Moreover, the test results are subject to test environments and the participants' preferences. Therefore, machine-based objective evaluations are used to provide a convenient, consistent, and fair assessment.

- Objective evaluation test

Objective evaluation test is intended to facilitate the implementation of subjective listening test. To achieve this goal, the results of objective evaluation should correlate well with the SDG scores.

Currently, the commonly used objective evaluation is to assess the perceptual quality of audio data via a stimulant ear, such as Evaluation of Audio Quality (EAQUAL) [47], Perceptual Evaluation of Audio Quality (PEAQ) [48], and Perceptual Model-Quality Assessment (PEMO-Q) [49]. Basically, these methods establish an auditory perception model to imitate the listening behavior of a human being, so that the watermarked signal is graded relative to the host signal. The whole process is depicted in Fig. 6.1 [31, 46]. After the watermark is embedded, the host and watermarked signals are separately passed to a psychoacoustic model. As described in Sect. 2.4, the psychoacoustic model calculates the internal representation of signal features, such as the masking threshold. By comparing the internal representations of the host and watermarked signals, the audible difference is determined. The audible difference is the input to the cognitive model, which models the cognitive processes in the human brain. After the audible difference is perceptually scaled in the cognitive model, the final output is an objective difference grade (ODG). As mentioned in Sect. 3.1.2, the specifications of ODG conform to those of SDG. To guarantee the accuracy of evaluation, a large set of relevant test signals are required to train and characterize such models [46].

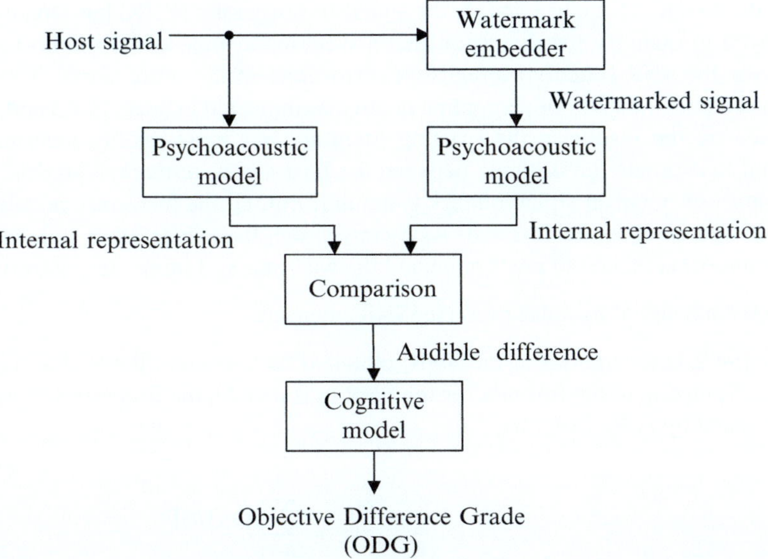

Fig. 6.1 Objective evaluation via perception modelling

Among the implemented models, PEMO-Q is the latest and most advanced predictor of audio quality. It is reported in [49] that PEMO-Q has a higher ability to be applicable to unknown distortions and performs better than the other techniques. The performance of three evaluation tools will be examined in Sect. 6.3.3.

Besides perception modelling, the extent of dissimilarity between the watermarked and host signals can be quantified by objective quality measures. Objective quality measures, such as the signal-to-noise ratio measure, the segmental signal-to-noise ratio measure, the cepstral distortion measure, the log-likelihood ratio measure, the Itakura–Saito distortion measure, the log-area ratio measure, and the weighted spectral slope measure [50], are commonly used in speech processing. They have been widely used in quality evaluation for speech enhancement [136–138], speech intelligibility estimation [139], speech recognition in blind source separation [140, 141], and noise reduction schemes [138]. We investigate using these quality measures for objective assessments of the perceptual quality of audio watermarking for the first time [142, 143].

6.2   Objective Quality Measures

Objective quality measures have been widely used in the quality evaluation of speech signals [50]. This kind of measurement makes use of sound source information and calculates the distance or distortion of the test signal with regard to the original signal [140], which corresponds to the concept of perceptual quality assessment in audio watermarking.

As discussed in Chaps. 3 and 5, the signal-to-noise ratio (SNR) has already been employed to quantify the distortion that a watermark imposes on the host signal. However, the SNR actually averages the distortions on the entire signal. Thus, it is not an accurate indicator of perceptual quality, as indicated in Sects. 3.2.5 and 5.2.2.

Based on the results in the existing literature, six more quality measures are selected to estimate the distance between the host and watermarked signals. Since the impact of noise on signal quality is nonuniform, all the measures calculate the level of distortion for each frame. As a convention, the subscripts o and w denote the components related to host frame and the watermarked frame, respectively.

- Segmental signal-to-noise ratio (segSNR) measure

The segSNR is a variation of the SNR, obtained by averaging the SNRs of all the frames. Referring to the formula for the SNR in Eq. (1.3), the frame-based segSNR is calculated by [136, 138, 140]

$$
d_{segSNR}\left(g_w,\, g_o\right) = 10\log_{10}\frac{\sum_{n=1}^{N}\left[g_o\left(n\right)\right]^2}{\sum_{n=1}^{N}\left[g_w\left(n\right) - g_o\left(n\right)\right]^2} \tag{6.1}
$$

where g_o is the host frame, g_w is the watermarked frame, and N is the frame length in samples. In our experiments, $N = 512$, which corresponds to 11.6 ms for a sampling rate of 44.1 kHz.

In fact, frames with segSNRs above 35 dB do not reflect human perceptual differences; therefore, their segSNRs are generally replaced with 35 dB. Moreover, silence frames have negative segSNRs because the signal energy is small. To prevent getting such abnormal segSNRs, a lower threshold for the segSNR is set to be -10 dB. Thus, the segSNR values are limited in the range of $[-10\,\text{dB},\, 35\,\text{dB}]$ [50, 136, 138].

- Cepstral distortion measure

The cepstral distortion (CD) measure provides an estimate of cepstral distance between the watermarked frame and the host frame. Given both cepstral coefficient vectors $\vec{c}_w$ and $\vec{c}_o$, CD for the first L coefficients is calculated by [140]

$$
d_{CD}\left(\vec{c}_w,\, \vec{c}_o\right) = \sum_{l=1}^{L}\left[\vec{c}_w\left(l\right) - \vec{c}_o\left(l\right)\right]^2 \tag{6.2}
$$

where $L = 50$ in our experiments.

- Log-likelihood ratio measure

The log-likelihood ratio (LLR or Itakura distance) measure is based on linear prediction (LP) analysis. Given both LP coefficient vectors $\vec{a}_w$ and $\vec{a}_o$, LLR measure is defined by [50, 136, 138, 140]

$$d_{LLR}\left(\vec{a}_w, \vec{a}_o\right) = \log_{10}\left(\frac{\vec{a}_w R_o \vec{a}_w^T}{\vec{a}_o R_o \vec{a}_o^T}\right) \tag{6.3}$$

where R_o is the autocorrelation matrix and $(\cdot)^T$ refers to the transpose of a matrix.

- Itakura–Saito distortion measure

The Itakura–Saito (IS) distortion measure is slightly different from the LLR measure and defined by [50, 138, 140]

$$d_{IS}\left(\vec{a}_m, \vec{a}_o\right) = \left(\frac{\sigma_o^2}{\sigma_w^2}\right) \cdot \left(\frac{\vec{a}_w R_o \vec{a}_w^T}{\vec{a}_o R_o \vec{a}_o^T}\right) + \log_{10}\left(\frac{\sigma_w^2}{\sigma_o^2}\right) - 1 \tag{6.4}$$

where σ_o^2 and σ_w^2 are all-pole gains for the host and watermarked frames, respectively.

It was mentioned in [140] that LLR and IS measures perform well as predictors of the recognition rate for the signals with additive noise in continuous speech recognition systems.

- Log-area ratio measure

The log-area ratio (LAR) measure is also based on LP analysis in that it depends on LP reflection coefficients [136–138, 140]:

$$d_{LAR}\left(\vec{r}_w, \vec{r}_o\right) = \left|\frac{1}{P}\sum_{p=1}^{P}\left[\log_{10}\frac{1+\vec{r}_o\left(p\right)}{1-\vec{r}_o\left(p\right)} - \log_{10}\frac{1+\vec{r}_w\left(p\right)}{1-\vec{r}_w\left(p\right)}\right]^2\right|^{1/2} \tag{6.5}$$

where P is the order of LP analysis and $P = 10$ in our experiments. $\vec{r}_o$ and $\vec{r}_w$ are the LP reflection coefficient vectors of the host and watermarked frames, respectively.

Since the reflection coefficients are closely related to power spectra, the LAR measure is able to estimate the differences between the logarithms of the spectra of the host and watermarked signals efficiently [137]. In [136, 137, 140], it has been observed that the LAR is the best measure in some cases.

- Weighted spectral slope measure

The Weighted spectral slope (WSS) measure is based on an auditory model, in which 36 overlapping filters of progressively larger bandwidth are used to estimate the smoothed short-time spectra [136]. Then, the weighted difference between the spectral slopes (SL) in each band are calculated [139].

According to [50, 136, 138, 140], the WSS measure in decibels is formulated as

$$d_{WSS} = K_{spl} (K_o - K_w) + \sum_{k=1}^{36} w_a (k) [SL_o (k) - SL_w (k)]^2 \qquad (6.6)$$

where K_o and K_w are related to the overall sound pressure level and K_{spl} is a parameter that can be varied to increase overall performance. In our experiments, $K_{spl} = 0$ is used as in [140] and the weight w_a depends on the formant locations [136]. As concluded in [50, 140], the WSS measure might outperform other measures because it employs the auditory model.

Note that for each objective quality measure (except segSNR), its overall quality score is obtained by using the $m_{95\%}$ mean to reduce the number of outliers. The $m_{95\%}$ mean of each quality measure is calculated in the following way. First, the value of the quality measure is calculated for each frame. Then the values of the quality measure for all the frames are sorted in an ascending order. The $m_{95\%}$ mean is the average of the first 95 % values of each quality measure [136, 138].

6.3 Experiments and Discussion

In this section, objective quality measures are evaluated to estimate their capabilities for predicting the perceptual quality of the watermarked audio signals. This is achieved by performing correlation analysis between the SDGs and the values of objective quality measures.

The audio signals used are taken from the test set prepared in Sect. 3.1.1, 17 pieces of audio signals ($A_1 \sim A_{17}$) in total. However, PEMO-Q is a commercial software tool and its demo version is strictly limited to signal lengths up to 4 s. Therefore, we always use a 4 s length from the beginning of each original audio test signal and then utilize them for the experiments here. Moreover, all the simulations are also conducted on a Pentium 4 2.4 GHz computer with 1 GB RAM under Windows XP operating system.

6.3.1 *Audio Watermarking Techniques Default Settings*

The performance of objective quality measures are fully investigated under different audio watermarking techniques, such as the proposed scheme in Chap. 4 along with cepstrum domain watermarking, wavelet domain watermarking, echo hiding, and histogram-based watermarking in Chap. 3.

In the experiments, each technique is employed to implement the process of watermarking separately. The imperceptibility of the watermarked signal is controlled by the watermark strength or a similar factor. Without loss of generality, all five watermarking techniques directly embed the pseudorandom number sequence (PRS) at full capacity. Based on the results in Chaps. 3 and 5, each watermarking technique uses the following parameters which have provided the best performance:

(1) Proposed audio watermarking scheme: frame length $N = 512$, the number of units per block $N_{unit} = 10$, the number of watermark bits embedded in one block $N_{bit} = 4$, the number of slots for embedding one watermark bit $N_B = 30$, the number of selected subbands $\tilde{N}_{subband} = 28$, and watermark length $N_w = 64$. In the experiments, every host signal is watermarked twenty times with a watermark strength of $\alpha_w = 10, 20, 30, \ldots, 200$.

(2) Cepstrum domain watermarking: frame length $N = 2048$, repetition coding $n_r = 3$, and watermark length $N_w = 57$. In the experiments, every host signal is watermarked eleven times with a watermark strength of $\alpha_w = 1 \times 10^{-3}, 1.2 \times 10^{-3}, 1.4 \times 10^{-3}, \ldots, 3 \times 10^{-3}$.

(3) Wavelet domain watermarking: frame length $N = 2048$, repetition coding $n_r = 3$, and watermark length $N_w = 57$. In the experiments, every host signal is watermarked ten times with a watermark strength of $\alpha_w = 0.01, 0.02, 0.03, \ldots, 0.1$.

(4) Echo hiding (kernel 3): frame length $N = 4096$, repetition coding $n_r = 3$, and watermark length $N_w = 28$. In the experiments, every host signal is watermarked eleven times with an echo amplitude of $\alpha = 0.1, 0.12, 0.14, \ldots, 0.3$.

(5) Histogram-based watermarking: the embedding strength $E_h = 1.4$ and watermark length $N_w = 10$. In the experiments, every host signal is watermarked eleven times with an embedding range of $\lambda = 2, 2.05, 2.1, \ldots, 2.5$.

Note that for each watermarking technique, every host signal has N_α watermarked signals: $N_\alpha = 20$ for the proposed scheme, $N_\alpha = 11$ for cepstrum domain watermarking, $N_\alpha = 10$ for wavelet domain watermarking, $N_\alpha = 11$ for echo hiding, and $N_\alpha = 11$ for histogram-based watermarking. The notation N_α will be used for correlation analysis to be undertaken in Sect. 6.3.4.

6.3.2 Subjective Listening Tests

Similar to the previous subjective listening tests, ten trained listeners participated in the tests that were performed in an isolated chamber. Also, all the stimuli were presented through a high-fidelity headphone.

During the tests, the participants were asked to evaluate the perceptual quality of the watermarked signal relative to its host signal and subsequently provide a SDG. In view of the difficulties in the real listening tests, only the proposed audio watermarking scheme was considered. Moreover, for each host signal, the human subjects were not required to evaluate all the twenty watermarked signals (i.e., $\alpha_w = 10, 20, 30, \ldots, 200$), but just five of them with $\alpha_w = 40, 80, 120, 160, 200$. In addition, the host signal was continuously included as a watermarked signal with $\alpha_w = 0$. This is because subjects are apt to make incorrect judgments in a situation where audible differences between test signals are too subtle to perceive. Therefore, for each host signal, there were six watermarked signals with $\alpha_w = 0, 40, 80, 120, 160, 200$.

In this way, every listener needed to participate in 17 separate tests on A_i, $i = 1, 2, \ldots, 17$. For each A_i, the six watermarked signals of A_i are denoted by $A_{ij'}$, $j' = 1, 2, \ldots, 6$. Since there were ten subjects participating in the tests, the SDG score of $A_{ij'}$ provided by the k-th subject is denoted by $G_{SDG}\left(i, j', k\right)$, where $k = 1, 2, \ldots, 10$. Then, the average SDG for host signal $A_{ij'}$ is calculated as

$$\tilde{G}_{SDG}\left(i, j'\right) = \frac{1}{K} \sum_{k=1}^{K} G_{SDG}\left(i, j', k\right) \tag{6.7}$$

where $K = 10$.

For simplicity of expression, the average SDGs for each host signal A_i is denoted as $\tilde{G}_{SDG}(i)$, where $\tilde{G}_{SDG}(i) = \left\{\tilde{G}_{SDG}\left(i, j'\right)\right\}$, $j' = 1, 2, \ldots, 6$.

6.3.3 Objective Evaluation Tests

Objective evaluation tests comprise two stages: investigation of the evaluation tools and calculation of the values of the quality measures.

- Evaluation tool analysis

In the first stage, we investigate the effectiveness of three evaluation tools using perception modelling, namely PEMO-Q [49], EAQUAL [47], and PEAQ [48]. The aim is to find the best quasi-subjective predictor of audio quality that would best conform to the SDG. Its ODGs will be adopted subsequently as quasi-SDGs for correlation analysis in the next section. The reason of using quasi-SDGs rather than SDGs is that it would be inaccurate to perform a correlation with an insufficient amount of the average SDGs.

To this purpose, all the watermarked signals of each host signal are evaluated separately using three evaluation tools.

Take the proposed audio watermarking scheme as an example. Each host signal has twenty watermarked signals with $\alpha_w = 10, 20, 30, \ldots, 200$. Moreover, the host signal is also included to correspond with its $\tilde{G}_{SDG}$ obtained above. So for each host signal A_i, there are twenty-one watermarked signals with $\alpha_w = 0, 10, 20, 30, \ldots, 200$. After being evaluated by three tools, each host signal A_i receives three kinds of ODGs, i.e., $G_{ODG1}(i) = \left\{G_{ODG1}\left(i, \hat{j}\right)\right\}$ by PEMO-Q, $G_{ODG2}(i) = \left\{G_{ODG2}\left(i, \hat{j}\right)\right\}$ by EAQUAL, and $G_{ODG3}(i) = \left\{G_{ODG3}\left(i, \hat{j}\right)\right\}$ by PEAQ, where $\hat{j} = 1, 2, \ldots, 21$.

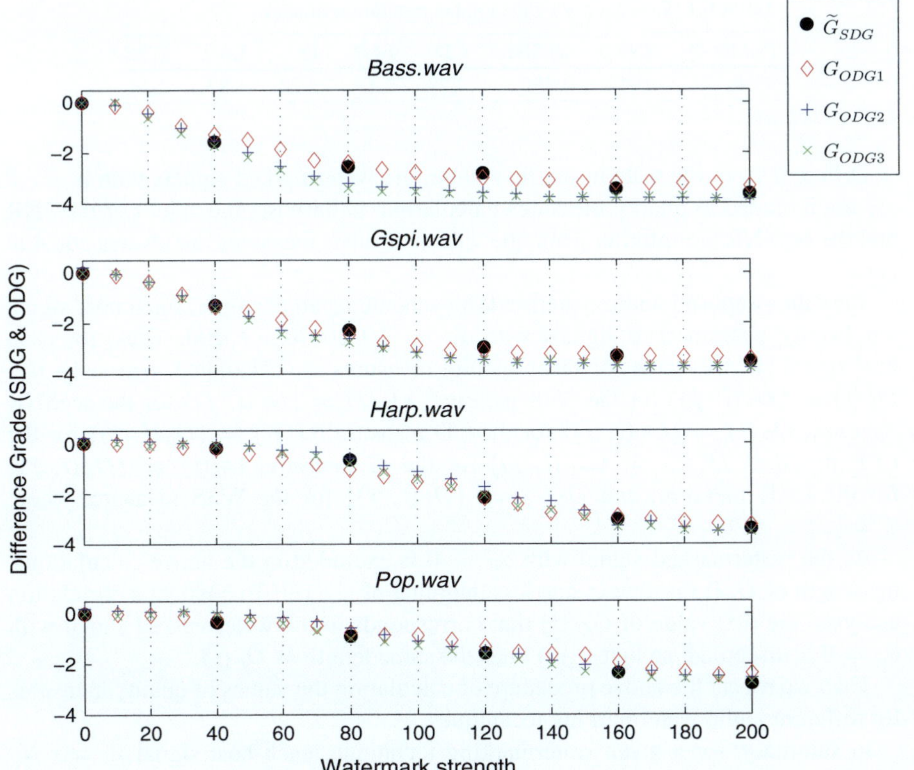

Fig. 6.2   Evaluation of PEMO-Q, PEAQ, and EAQUAL

Figure 6.2 shows the average SDGs and the ODGs for the watermarked *Bass.wav*, *Gspi.wav*, *Harp.wav*, and *Pop.wav* signals, respectively.

Note that a few ODGs in Fig. 6.2 are slightly positive, as with some values with small watermark strengths. As mentioned in Sect. 3.2.1, such cases are interpreted as distortions that are mostly inaudible for humans.

According to Fig. 6.2, for each host signal, its G_{ODG1} are closer to $\tilde{G}_{SDG}$ than G_{ODG2} and G_{ODG3}. It means that PEMO-Q provides a better correspondence between subjective and objective difference grades. Therefore $G_{ODG1}(i)$ or simplified as $G_Q(i)$ are used for correlation analysis.

- Quality measures calculation

In the second stage, we calculate the values of the quality measures between the host signal and all its watermarked signals. The selected objective quality measures include the SNR, segSNR, CD, LLR, IS, LAR, and WSS measures, denoted by $r = 1, 2, \ldots, 7$, respectively. Consequently, each host signal A_i has the values of seven quality measures, denoted by $O_r(i)$, $r = 1, 2, \ldots, 7$.

Table 6.1 Comparison of the total computation time (s)

PEMO-Q	SNR	segSNR	CD	LLR	IS	LAR	WSS
55.48	0.012	0.051	3.78	2.19	2.20	1.82	2.24

Different from the tests in the first stage, the watermarked signals with $\alpha_w = 0$ are not included in quality measures calculation. Otherwise, the values of the SNR and the segSNR are infinite, while the values of other measures are always equal to zero.

Take the proposed audio watermarking scheme as an example. Each host signal has twenty watermarked signals with $\alpha_w = 10, 20, 30, \ldots, 200$. Thus, for each host signal A_i, the values of seven quality measures are calculated separately, i.e., $O_1(i) = \{O_1(i, j)\}$ for the SNR measure, $O_2(i) = \{O_2(i, j)\}$ for the segSNR measure, $O_3(i) = \{O_3(i, j)\}$ for the CD measure, $O_4(i) = \{O_4(i, j)\}$ for the LLR measure, $O_5(i) = \{O_5(i, j)\}$ for the IS measure, $O_6(i) = \{O_6(i, j)\}$ for the LAR measure, and $O_7(i) = \{O_7(i, j)\}$ for the WSS measure, where $j = 1, 2, \ldots, 20$.

As the watermarked signal with $\alpha_w = 0$ is excluded in the above calculations, the length of $O_r(i)$ is always one less than that of $G_Q(i)$. To conduct a correlation analysis, the first value of $G_Q(i)$ that corresponds to the watermarked signal with $\alpha_w = 0$ is discarded, so that $G_Q(i)$ has the same length as $O_r(i)$.

Then we repeat the above procedure of calculating the values of quality measures for different audio watermarking techniques.

In summary, for a given watermarking technique, each host signal A_i has N_α watermarked signals, as introduced in Sect. 6.3.1. Based on these N_α watermarked signals, each A_i receives the quasi-SDGs $G_Q(i) = \{G_Q(i, j)\}$ and the values of seven quality measures $O_r(i) = \{O_r(i, j)\}$, $r = 1, 2, \ldots, 7$, where $j = 1, 2, \ldots, N_\alpha$.

Note that computation time is also one of our concerns. Table 6.1 lists the computation time of quality measures on one watermarked *Bass.wav* signal with $\alpha_w = 60$ in the proposed audio watermarking scheme. PEMO-Q took around 55 s to complete the evaluation of one watermarked signal with the default settings in [49]. In comparison, all quality measures finished in less than 4 s, much faster than PEMO-Q. Particularly the SNR and segSNR measures took the least time, less than 0.1 s. Also, the computation time of the LAR, LLR, and IS measures are not more than 2.2 s. The measured response times are based on using Pentium 4 PC empowered by 2.4 GHz CPU and 1 GB RAM running under Windows XP operating system.

6.3.4 Performance Evaluation Using Correlation Analysis

To evaluate the performance of objective quality measures, Pearson correlation coefficient ρ is calculated between the values of each quality measure, O_r, and its quasi-SDGs, G_Q. Commonly, ρ is defined by [138–140]

$$\rho = \frac{\sum_{n=1}^{N_m} \left[O_r(n) - \overline{O}_r\right]\left[G_Q(n) - \overline{G}_Q\right]}{\left\{\sum_{n=1}^{N_m} \left(O_r(n) - \overline{O}_r\right)^2\right\}^{1/2} \left\{\sum_{n=1}^{N_m} \left(G_Q(n) - \overline{G}_Q\right)^2\right\}^{1/2}} \tag{6.8}$$

where N_m is the length of O_r and G_Q. $\overline{O}_r$ and $\overline{G}_Q$ are the means of O_r and G_Q, respectively.

Note that a correlation coefficient is a number between -1 and 1. If the coefficient is closer to 1 (positive correlation) or -1 (inverse correlation), it indicates that the values of quality measure are in higher correlation with the quasi-SDGs. If the coefficient is closer to 0, it indicates that there is less correlation between the values of quality measure and the quasi-SDGs.

For each audio watermarking technique, two types of correlation analyses are conducted [138]. Recall that the indices i, j, r are for indicating the host signal, watermark strength, and quality measure, respectively.

In the first analysis, the correlation is separately performed on each host signal. Given a host signal A_i, its individual correlation coefficient with the rth objective quality measure $\rho(i, r)$, $i = 1, 2, \ldots, 17$ and $r = 1, 2, \ldots, 7$ is calculated by [143]

$$\rho(i, r) = \frac{\sum_{j=1}^{N_\alpha} \left[O_r(i, j) - \overline{O}_r(i)\right]\left[G_Q(i, j) - \overline{G}_Q(i)\right]}{\left\{\sum_{j=1}^{N_\alpha} \left[O_r(i, j) - \overline{O}_r(i)\right]^2\right\}^{1/2} \left\{\sum_{j=1}^{N_\alpha} \left[G_Q(i, j) - \overline{G}_Q(i)\right]^2\right\}^{1/2}} \tag{6.9}$$

where $\overline{O}_r(i) = \frac{1}{N_\alpha}\sum_{j=1}^{N_\alpha} O_r(i, j)$ and $\overline{G}_Q(i) = \frac{1}{N_\alpha}\sum_{j=1}^{N_\alpha} G_Q(i, j)$.

The average correlation coefficient of each quality measure $\rho_1(r)$, $r = 1, 2, \ldots, 7$ is calculated by

$$\rho_1(r) = \frac{1}{N_h}\sum_{i=1}^{N_h} \rho(i, r) \tag{6.10}$$

where $N_h = 17$ is the number of host signals used in the experiments.

In the second analysis, the correlation is directly performed on all the host signals. The overall correlation coefficient of the rth objective quality measure $\rho_2(r)$, $r = 1, 2, \ldots, 7$ is calculated by [143]

$$\rho_2(r) = \frac{\sum_{i=1}^{N_h}\sum_{j=1}^{N_\alpha}[O_r(i,j) - \overline{O}_r][G_Q(i,j) - \overline{G}_Q]}{\left\{\sum_{i=1}^{N_h}\sum_{j=1}^{N_\alpha}[O_r(i,j) - \overline{O}_r]^2\right\}^{1/2}\left\{\sum_{i=1}^{N_h}\sum_{j=1}^{N_\alpha}[G_Q(i,j) - \overline{G}_Q]^2\right\}^{1/2}} \tag{6.11}$$

where $\overline{O}_r = \frac{1}{N_h \cdot N_a}\sum_{i=1}^{N_h}\sum_{j=1}^{N_\alpha}O_r(i,j)$ and $\overline{G}_Q = \frac{1}{N_h \cdot N_a}\sum_{i=1}^{N_h}\sum_{j=1}^{N_\alpha}G_Q(i,j)$.

Note that the average correlation coefficient is widely used in studying objective quality measures [136–141]. The overall correlation coefficient is more desirable in some applications, but considered to be rather stringent [138, 144].

Tables 6.2, 6.3, 6.4, 6.5, and 6.6 show the Pearson correlation coefficients under different audio watermarking techniques. The results include the individual correlation coefficients $\rho(i, r)$, the average correlation coefficients (absolute value) $|\rho_1(r)|$, and the overall correlation coefficients (absolute value) $|\rho_2(r)|$, where $i = 1, 2, \ldots, 17$ for denoting the host signal A_i and $r = 1, 2, \ldots, 7$ for denoting the quality measure. In each table, the highest $|\rho_1(r)|$ and $|\rho_2(r)|$ (i.e., the absolute value closer to 1) are shaded and the second highest ones are in bold.

Some observations can be obtained from the Pearson correlation coefficients [143].

- The overall correlation coefficients $|\rho_2(r)|$ are generally lower than the average correlation coefficients $|\rho_1(r)|$ under different audio watermarking techniques.

  Take the proposed audio watermarking scheme in Table 6.2 as an example. $|\rho_1(r)|$, $r = 1, 2, \ldots, 7$ are equal to 0.92, 0.87, 0.92, 0.95, 0.85, 0.95, and 0.94, respectively, not less than 0.85. However, $|\rho_2(r)|$, $r = 1, 2, \ldots, 7$ are equal to 0.38, 0.30, 0.26, 0.58, 0.27, 0.64, and 0.59, respectively, not more than 0.64.

  This is because the functional relationship between objective quality measure and the quasi-SDGs varies across different types of audio signals, i.e., vocal, percussive instrument, tonal instrument, and music. Even in the same category, different instruments or different genres of music are most likely to exhibit different time–frequency characteristics. Consequently, the overall correlation coefficients $|\rho_2(r)|$ are less than average correlation coefficients $|\rho_1(r)|$ in most cases of audio watermarking techniques, whereas exceptions exist, due to the intricacy of different techniques.

  If audio signals have similar properties, the overall correlation coefficients become better. For instance, host audio signals A_1 (*Soprano.wav*), A_2 (*Bass.wav*), and A_3 (*Quartet.wav*) all belong to the vocal category and also have the same lyrics. Then the overall correlation coefficients over A_1, A_2, and A_3 can be calculated by eq. (6.11), where $N_h = 3$. Figure 6.3 shows the results of $\rho_2(r)$

Table 6.2 Pearson correlation coefficients under our proposed scheme

		SNR $r=1$	segSNR $r=2$	CD $r=3$	LLR $r=4$	IS $r=5$	LAR $r=6$	WSS $r=7$
Individual	$A_{i=1}$	0.97	0.86	−0.97	−0.98	−0.86	−0.98	−0.93
coeff.	$A_{i=2}$	0.98	0.95	−0.93	−0.99	−0.79	−0.99	−0.96
$\rho(i, r)$	$A_{i=3}$	0.92	0.99	−0.99	−0.94	−0.93	−0.95	−0.98
	$A_{i=4}$	0.93	0.93	−0.95	−0.96	−0.89	−0.98	−0.96
	$A_{i=5}$	0.89	0.95	−0.98	−0.96	−0.97	−0.93	−0.99
	$A_{i=6}$	0.72	0.36	−0.47	−0.71	−0.33	−0.72	−0.57
	$A_{i=7}$	0.98	0.91	−0.88	−0.99	−0.78	−0.98	−0.90
	$A_{i=8}$	0.91	0.97	−0.97	−0.99	−0.94	−0.96	−0.96
	$A_{i=9}$	0.93	0.98	−0.97	−0.98	−0.90	−0.97	−0.93
	$A_{i=10}$	0.97	0.64	−0.90	−0.98	−0.74	−0.98	−0.96
	$A_{i=11}$	0.93	0.98	−0.98	−0.96	−0.94	−0.93	−0.97
	$A_{i=12}$	0.97	0.94	−0.94	−0.97	−0.82	−0.96	−0.98
	$A_{i=13}$	0.92	0.50	−0.81	−0.97	−0.60	−0.94	−0.89
	$A_{i=14}$	0.87	0.98	−0.97	−0.85	−0.98	−0.94	−0.99
	$A_{i=15}$	0.91	0.99	−0.99	−0.99	−0.94	−0.98	−0.98
	$A_{i=16}$	0.89	0.98	−0.98	−0.97	−0.98	−0.94	−0.99
	$A_{i=17}$	0.88	0.96	−0.97	−0.98	−0.97	−0.94	−0.99
Average coeff. $\lvert\rho_1(r)\rvert$		0.92	0.87	**0.92**	0.95	0.85	0.95	0.94
Overall coeff. $\lvert\rho_2(r)\rvert$		0.38	0.30	0.26	0.58	0.27	0.64	**0.59**

Table 6.3 Pearson correlation coefficients under cepstrum domain watermarking

		SNR $r=1$	segSNR $r=2$	CD $r=3$	LLR $r=4$	IS $r=5$	LAR $r=6$	WSS $r=7$
Individual	$A_{i=1}$	0.82	0.78	−0.62	−0.93	−0.87	−0.92	−0.90
coeff.	$A_{i=2}$	0.84	0.83	−0.88	−0.99	−0.99	−0.99	−0.81
$\rho(i, r)$	$A_{i=3}$	0.80	0.79	−0.87	−1.00	−0.99	−0.98	−0.79
	$A_{i=4}$	0.94	0.95	−0.85	−0.94	−0.56	−0.96	−0.97
	$A_{i=5}$	0.81	0.68	−0.87	−0.95	−0.99	−0.89	−0.76
	$A_{i=6}$	0.72	0.92	−0.82	−0.81	−0.43	−0.89	−0.83
	$A_{i=7}$	0.96	0.96	−0.93	−0.91	−0.74	−0.98	−0.98
	$A_{i=8}$	0.74	0.66	−0.84	−0.98	−0.99	−0.90	−0.86
	$A_{i=9}$	0.79	0.70	−0.80	−1.00	−1.00	−0.92	−0.91
	$A_{i=10}$	0.92	0.90	−0.89	−0.90	−0.84	−0.98	−0.97
	$A_{i=11}$	0.46	0.39	−0.38	−0.78	−0.90	−0.67	−0.64
	$A_{i=12}$	0.96	0.95	−0.98	−0.91	−0.81	−0.98	−0.98
	$A_{i=13}$	0.91	0.94	−0.57	−0.77	−0.41	−0.85	−0.71
	$A_{i=14}$	0.72	0.68	−0.82	−0.99	−0.99	−0.93	−0.76
	$A_{i=15}$	0.81	0.77	−0.77	−1.00	−1.00	−0.95	−0.85
	$A_{i=16}$	0.81	0.77	−0.82	−1.00	−1.00	−0.96	−0.84
	$A_{i=17}$	−0.23	−0.31	0.20	0.12	0.14	0.11	0.26
Average coeff. $\lvert\rho_1(r)\rvert$		0.75	0.73	0.73	0.87	0.79	**0.86**	0.78
Overall coeff. $\lvert\rho_2(r)\rvert$		0.36	0.39	0.17	0.72	0.37	**0.70**	0.49

Table 6.4 Pearson correlation coefficients under wavelet domain watermarking

		SNR $r=1$	segSNR $r=2$	CD $r=3$	LLR $r=4$	IS $r=5$	LAR $r=6$	WSS $r=7$
Individual	$A_{i=1}$	0.86	0.84	−0.77	−1.00	−1.00	−0.99	−0.96
coeff.	$A_{i=2}$	0.86	0.84	−0.88	−0.99	−1.00	−0.99	−0.89
$\rho(i, r)$	$A_{i=3}$	0.93	0.91	−0.84	−0.97	−0.98	−1.00	−0.93
	$A_{i=4}$	0.96	0.97	−0.79	−0.94	−0.71	−0.96	−0.93
	$A_{i=5}$	0.85	0.79	−0.90	−0.99	−0.99	−0.96	−0.82
	$A_{i=6}$	0.74	0.91	−0.90	−0.82	−0.60	−0.90	−0.83
	$A_{i=7}$	0.98	0.97	−0.90	−0.90	−0.87	−0.97	−0.99
	$A_{i=8}$	0.74	0.70	−0.78	−0.97	−0.91	−0.93	−0.79
	$A_{i=9}$	0.64	0.62	−0.68	−0.87	−0.74	−0.82	−0.76
	$A_{i=10}$	0.89	0.86	−0.82	−0.98	−0.69	−1.00	−0.97
	$A_{i=11}$	−0.83	−0.81	0.78	0.94	0.94	0.91	0.88
	$A_{i=12}$	0.94	0.93	−0.94	−0.97	−0.98	−1.00	−0.98
	$A_{i=13}$	0.90	0.93	0.76	−0.76	−0.75	−0.88	−0.44
	$A_{i=14}$	−0.11	−0.15	0.09	−0.42	−0.35	−0.27	0.18
	$A_{i=15}$	0.87	0.84	−0.34	−0.45	−0.44	−0.40	−0.32
	$A_{i=16}$	0.81	0.78	−0.78	−0.98	−0.98	−0.90	−0.87
	$A_{i=17}$	0.70	0.65	−0.69	−0.91	−0.85	−0.81	−0.75
Average coeff. $|\rho_1(r)|$		0.69	0.68	0.55	**0.76**	**0.70**	0.76	0.66
Overall coeff. $|\rho_2(r)|$		0.14	0.31	0.18	**0.68**	0.44	0.71	0.43

Table 6.5 Pearson correlation coefficients under echo hiding

		SNR $r=1$	segSNR $r=2$	CD $r=3$	LLR $r=4$	IS $r=5$	LAR $r=6$	WSS $r=7$
Individual	$A_{i=1}$	0.96	0.96	−0.96	−1.00	−1.00	−0.99	−1.00
coeff.	$A_{i=2}$	0.99	0.99	−0.99	−1.00	−0.98	−1.00	−0.99
$\rho(i, r)$	$A_{i=3}$	0.97	0.97	−0.99	−1.00	−0.97	−0.99	−1.00
	$A_{i=4}$	0.99	0.99	−0.99	−1.00	−0.99	−1.00	−0.99
	$A_{i=5}$	0.98	0.97	−0.99	−1.00	−1.00	−0.99	−1.00
	$A_{i=6}$	0.93	0.98	−0.91	−0.85	−0.84	−0.90	−0.83
	$A_{i=7}$	0.94	0.94	−0.99	−1.00	−0.99	−0.99	−1.00
	$A_{i=8}$	0.98	0.98	−0.99	−0.99	−1.00	−0.99	−0.99
	$A_{i=9}$	0.96	0.96	−0.95	−1.00	−0.98	−0.98	−1.00
	$A_{i=10}$	0.94	0.93	−0.96	−0.99	−0.96	−0.98	−0.99
	$A_{i=11}$	0.99	0.99	−1.00	−0.98	−0.98	−1.00	−0.98
	$A_{i=12}$	0.95	0.96	−0.98	−1.00	−0.99	−0.99	−1.00
	$A_{i=13}$	0.02	0.16	−0.13	−0.19	−0.14	−0.17	−0.12
	$A_{i=14}$	0.71	0.71	−0.72	−0.79	−0.80	−0.77	−0.79
	$A_{i=15}$	0.97	0.97	−0.97	−1.00	−1.00	−0.99	−1.00
	$A_{i=16}$	0.97	0.97	−0.99	−1.00	−0.99	−0.99	−1.00
	$A_{i=17}$	0.97	0.96	−0.99	−1.00	−1.00	−0.99	−1.00
Average coeff. $|\rho_1(r)|$		0.90	0.90	0.91	0.93	**0.92**	**0.92**	**0.92**
Overall coeff. $|\rho_2(r)|$		0.53	0.32	0.05	**0.59**	0.39	0.65	0.49

Table 6.6 Pearson correlation coefficients under histogram-based watermarking

		SNR $r=1$	segSNR $r=2$	CD $r=3$	LLR $r=4$	IS $r=5$	LAR $r=6$	WSS $r=7$
Individual	$A_{i=1}$	0.76	0.81	−0.82	−0.73	−0.79	−0.75	−0.66
coeff.	$A_{i=2}$	0.83	0.82	0.38	0.67	−0.79	0.47	−0.71
$\rho(i, r)$	$A_{i=3}$	0.74	0.67	−0.54	−0.58	−0.72	−0.64	−0.76
	$A_{i=4}$	−0.12	−0.24	0.23	−0.07	−0.08	−0.07	0.47
	$A_{i=5}$	0.24	0.66	−0.01	−0.40	−0.24	−0.40	−0.35
	$A_{i=6}$	0.94	0.54	−0.89	−0.91	−0.85	−0.89	−0.91
	$A_{i=7}$	0.90	0.73	−0.33	−0.72	−0.83	−0.73	−0.88
	$A_{i=8}$	0.87	0.33	−0.64	−0.79	−0.81	−0.60	−0.83
	$A_{i=9}$	0.49	0.20	−0.59	−0.36	−0.52	−0.38	−0.33
	$A_{i=10}$	0.72	0.45	−0.65	−0.11	−0.66	−0.18	0.14
	$A_{i=11}$	0.85	0.17	−0.82	−0.75	−0.84	−0.75	−0.81
	$A_{i=12}$	−0.38	−0.42	−0.60	0.53	0.39	0.37	0.20
	$A_{i=13}$	0.07	0.07	0.10	0.10	−0.08	0.35	0.74
	$A_{i=14}$	0.95	0.43	−0.87	−0.91	−0.93	−0.91	−0.86
	$A_{i=15}$	0.64	0.79	−0.31	−0.20	−0.55	0.04	−0.86
	$A_{i=16}$	0.98	0.91	−0.47	−0.98	−0.92	−0.71	−0.93
	$A_{i=17}$	0.95	0.94	−0.76	−0.84	−0.87	−0.86	−0.88
Average coeff. $|\rho_1(r)|$		0.61	0.46	0.45	0.41	**0.59**	0.39	0.48
Overall coeff. $|\rho_2(r)|$		0.40	0.43	0.28	0.35	0.22	0.71	**0.50**

over A_1, A_2, and A_3 in the proposed audio watermarking scheme. It can be seen that $|\rho_2(r)|$, $r = 1, 2, \ldots, 7$ increase greatly to 0.84, 0.79, 0.85, 0.94, 0.81, 0.96, and 0.94 respectively, not less than 0.79.

- Under different watermarking techniques, the LAR measure ($r = 6$) shows the best performance in both overall and average correlations. The LAR measure provides the highest overall correlation under the proposed audio watermarking scheme ($|\rho_2(6)| = 0.64$), wavelet domain watermarking ($|\rho_2(6)| = 0.71$), echo hiding ($|\rho_2(6)| = 0.65$), and histogram-based watermarking ($|\rho_2(6)| = 0.71$). Under cepstrum domain watermarking, the LAR measure yields the second highest overall correlation ($|\rho_2(6)| = 0.70$), only slightly less than the highest value. Moreover, the LAR measure provides the highest average correlation under the proposed scheme ($|\rho_1(6)| = 0.95$) and wavelet domain watermarking ($|\rho_1(6)| = 0.76$). Also, the LAR measure yields the second highest average correlation under cepstrum domain watermarking ($|\rho_1(6)| = 0.86$) and echo hiding ($|\rho_1(6)| = 0.92$). However, the LAR measure receives the lowest average correlation under histogram-based watermarking ($|\rho_1(6)| = 0.39$).

  After the LAR measure, the LLR measure ($r = 4$) is also a good measure under different watermarking techniques. The LLR measure provides the highest overall correlation under cepstrum domain watermarking ($|\rho_2(4)| = 0.72$). Also, the LLR measure provides the second highest overall correlation under wavelet domain watermarking ($|\rho_2(4)| = 0.68$) and echo hiding

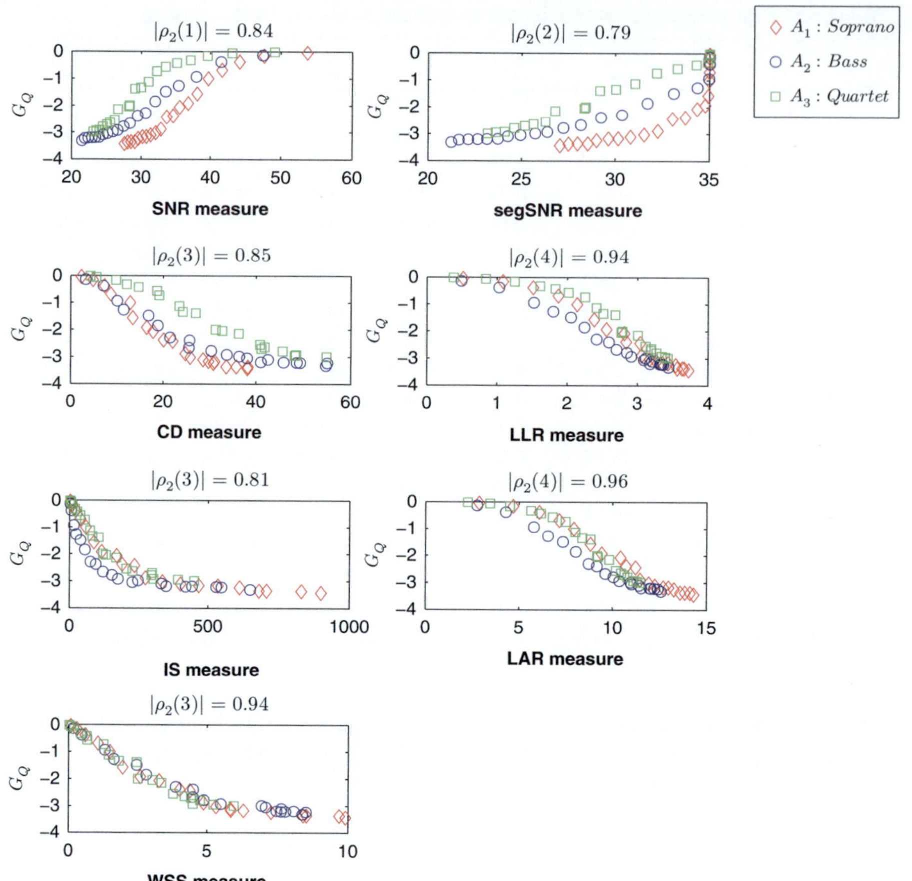

Fig. 6.3 Overall correlation coefficients over audio test signals A_1, A_2, and A_3

($|\rho_2(4)| = 0.59$). Moreover, the LLR measure provides the highest average correlation under the proposed scheme ($|\rho_1(4)| = 0.95$), cepstrum domain watermarking ($|\rho_1(4)| = 0.87$), wavelet domain watermarking ($|\rho_1(4)| = 0.76$), and echo hiding ($|\rho_1(4)| = 0.93$). However, the LLR measure receives the second lowest average correlation under histogram-based watermarking ($|\rho_1(4)| = 0.41$).

In addition, the WSS measure ($r = 7$) shows similar performance to the IS measure ($r = 5$), better than the SNR measure ($r = 1$) and the segSNR measure ($r = 2$) on the whole.

By comparison, the CD measure ($r = 3$) yields the worst correlation in most cases, especially quite low overall correlation. The CD measure yields the lowest overall correlation under the proposed scheme ($|\rho_2(3)| = 0.26$), cepstrum domain watermarking ($|\rho_2(3)| = 0.17$), and echo hiding ($|\rho_2(3)| = 0.05$).

Also, the CD measure yields the second lowest overall correlation under wavelet domain watermarking ($|\rho_2 (3)| = 0.18$) and histogram-based watermarking ($|\rho_2 (3)| = 0.28$).

- By using different quality measures, every audio watermarking technique can achieve a satisfactory overall correlation. The highest overall correlation coefficient is equal to 0.64 under the proposed scheme, equal to 0.72 under cepstrum domain watermarking, equal to 0.71 under wavelet domain watermarking, equal to 0.65 under echo hiding, and equal to 0.71 under histogram-based watermarking. As mentioned above, except for the fact that the highest overall correlation under cepstrum domain watermarking is provided by the LLR measure, the highest overall correlation under other watermarking techniques is provided by the LAR measure.

  This proves that objective quality measures are able to predict the perceptual quality of the watermarked audio signals.

6.4 Summary

Imperceptibility is one prime concern in audio watermarking for copyrights protection. In this chapter, objective quality measures used in speech quality evaluation have been assessed for their capabilities in the perceptual quality evaluation of audio watermarking.

Different from perception modelling that mimics the human auditory system, objective quality measures are adopted as an alternative approach to quantify the dissimilarities caused by audio watermarking in an objective manner. Various audio watermarking techniques discussed in Chaps. 3 and 4, such as our proposed scheme, cepstrum domain watermarking, wavelet domain watermarking, echo hiding, and histogram-based watermarking, are taken into consideration. During the experiments using each technique, subjective listening tests and a commercial evaluation tool PEMO-Q are used to grade the watermarked signals with different watermark strengths. Moreover, the distances between the watermarked and host signals are quantified by seven well-developed quality measures, i.e., the SNR, segSNR, CD, LLR, IS, LAR, and WSS measures. Then two types of Pearson correlation analyses are conducted to evaluate the performance of these quality measures serving as the predictors for perceptual quality. For each quality measure, one analysis is to calculate the average correlation coefficient by averaging the corresponding individual correlation coefficients over different test signals, and the other analysis is to calculate the overall correlation coefficient.

Pearson correlation coefficients show that the overall correlation coefficients are commonly lower than the corresponding average correlation coefficients. It is because the audio signals under test belong to different categories and possess different time–frequency characteristics. Nevertheless, the investigated quality measures, especially the LAR and the LLR measures, correlate well with the quasi-SDGs from PEMO-Q. Moreover, quality measures run much faster than PEMO-Q. These experimental results indicate that objective quality measures can be used reliably to estimate the perceptual quality of the watermarked audio signals.

Appendix A
SDMI Standard

Secure digital music initiative (SDMI) is aiming at developing open technology specifications that protect playing, storing, and distributing of digital music. To achieve the goal of copyright enforcement, SDMI embeds robust and secure watermarks (specifically called screening control data, SCD) into the music. Then SDMI-compliant devices which are fitted with watermark detectors [145] can identify the status of watermarks to perform appropriate operations.

In [146], key technical factors for evaluation were set forth in accordance with the claimed performance of the technology, including inaudibility, robustness, reliability, renewability, efficiency of operation, and effect on ability to compress content. First, the requirement for inaudibility is that the content containing SCD should be perceived as being statistically indistinguishable from the content prior to the addition of SCD. For the robustness, the watermarked content must be able to withstand each signal process listed in Table A.1. Here, the content is supposed to be sampled at 44.1 or 48 kHz and quantized at 16 bits. Moreover, false-negative and false-positive probabilities[1] for reliability are required to be no less than 10^{-2} and 10^{-12}, respectively. Third, renewability is expected to some extent, which indicates that the technology should fail in limited ways after a successful attack and also should provide a reasonable method of recovering from systematic compromise. In addition, the technology must operate on a number of platforms to estimate the efficiency of operation, i.e., measuring the amount of time necessary to detect or embed SCD in relation to the length of the excerpt. Finally, the technology should not interfere with the ability of standard compression algorithms to maintain an expected fidelity level at standard bit rates.

---

[1] False-negative probability is defined as the probability of missing detecting the existed watermark and false-positive probability is defined as the probability of detecting the nonexisted watermark.

Y. Lin and W.H. Abdulla, *Audio Watermark: A Comprehensive Foundation Using MATLAB*, 177
DOI 10.1007/978-3-319-07974-5, © Springer International Publishing Switzerland 2015

Table A.1 Robustness test items in SDMI

Signal process	Description
D/A, A/D	D/A, A/D, converting twice
Equalization	Typical case: 10-band graphic equalizer with the following characteristics
	Freq./Hz: 31 62 125 250 500 1 k 2 k 4 k 8 k 16 k
	Gain/dB: -6 $+6$ -6 $+6$ -6 $+6$ -6 $+6$ -6 $+6$
Band-pass filtering	100–6 kHz, 12 dB/oct.
Linear speed change	$\pm 10\,\%$
Codecs (at typically used data rates)	ISO/IEC 13818-7: 1997 ("AAC")
	ISO/IEC 14496-3: 1999 (MPEG-4 AAC with perceptual noise substitution)
	ISO/IEC 11172-3: 1993 Layer III (MPEG-1 Audio Layer 3 "MP3")
	Q-Design
	Windows Media Audio
	Twin-VQ
	ATRAC-3
	Dolby Digital AC-3 ATSC A_52
	ePAC
Noise addition	Adding white noise with constant level of 36 dB lower than total averaged music power (S/N: 36 dB)
Time-scale modification	Pitch-invariant time scaling: $\pm 4\,\%$
Wow and flutter	0.5 % rms, from DC to 250 Hz
Addition echo	Maximum delay: 100 ms
	Feedback coefficient: up to 0.5
Down mixing and surround sound processing	6-channel to stereo
	SRS
	Spatializer
	Dolby Surround
	Dolby Headphone
Sample rate conversion	48 kHz $\rightarrow$ 44.1 kHz
	96 kHz $\rightarrow$ 48/44.1 kHz
Dynamic range reduction	Threshold: 50 dB
	16 dB max compression
	Rate: 10 ms attack, 3 s recovery

Appendix B
STEP 2000

STEP 2000 [40] is a joint international evaluation project for audio digital watermarking technology, undertaken by JASRAC[1] and NRI[2] together with international associations of copyright management societies, CISAC and BIEM. It is the first work of its kind initiated by copyright management bodies.

The objective of STEP 2000 is "to certify the aptitude of digital watermark technologies, with a view towards promoting its utilization." Enthusiastic responses from many technology enterprises were received, contributing to an extensive technology evaluation.

The evaluation of submitted digital watermark technologies was conducted mainly with two aspects, i.e., audibility and robustness.

- *Audibility*—Whether the professionals can perceive if watermarks have been embedded in music that is played back in a recording studio environment

Subjective listening test, ABX test, was conducted in perceptual quality evaluation. First, the listener listens to a sound recording with no watermark (A), a sound recording with watermarks embedded (B), and a sound recording which is one of the two (X). After that, a listener listens to A and B alternately twice for 40 s each and listens to X for 40 s again. Then, the listener decides whether X is A or B.

There are two requirements in ABX test to ensure its validity. One is to eliminate (correct) contingency responses. To this end, the above tests were conducted five times for each system. Moreover, the listener is defined to have detected the embedded watermark if the same listener correctly determines whether the watermark is embedded or not on each of the five tests. Under this definition, significance of the responses are 95 % or greater. The other is to ensure typicality of the professionals from the recording industry. For this purpose, a group comprising of one recording engineer, one mastering engineer, one synthesizer manipulator, and one audio critic was selected.

---

[1]JASRAC: Japanese Society for Rights of Authors, Composers and Publishers
[2]NRI: Nomura Research Institute, Ltd.

Y. Lin and W.H. Abdulla, *Audio Watermark: A Comprehensive Foundation Using MATLAB*, 179
DOI 10.1007/978-3-319-07974-5, © Springer International Publishing Switzerland 2015

Table B.1 Robustness test items in STEP 2000

Testing item	Overview of processing involved
D/A, A/D transition	Digital→Analog→Digital
Altered number of channels	Stereo (2ch)→mono
Down sampling	44.1 kHz/16 bit/2ch→16 kHz/16 bit/2ch
Amplitude compression	44.1 kHz/16 bit/2ch→44.1 kHz/8 bit/2ch
Time and pitch compression and decompression	Time compression/decompression: $\pm 10\,\%$ Pitch shift compression/decompression: $\pm 10\,\%$
Linear data compression	MPEG 1 Audio Layer 3 (MP3): 128 kbps MPEG 2 AAC: 128 kbps ATRAC: Version 4.5 ATRAC 3: 105 kbps RealAudio: ISDN Windows Media Audio: ISDN
Nonlinear data compression	FM (FM multiple broadcast, terrestrial hertzian TV broadcast) AM (AM broadcast) PCM (Satellite TV broadcast: communications satellite, broadcasting satellite)
Characteristic transformation of frequency response	FM (FM multiple broadcast, terrestrial hertzian TV broadcast) AM (AM broadcast) PCM (Satellite TV broadcast: communications satellite, broadcasting satellite)
Noise	White noise: $S/N = -40\,\mathrm{dB}$

- *Robustness*—Whether the watermarked data can be extracted after various processes of music usage

The robustness tests were performed under equal conditions for all the submitted technologies. In general, watermarked data were manipulated in some way, for example, processed in a mastering studio, processed in a broadcasting studio (and a prospective broadcasting environment), processed for distribution through the Internet and other networks, and processed by commonly available consumer level equipments. Say specifically, robustness testing items are listed in Table B.1.

Appendix C
StirMark for Audio

StirMark for Audio [134] is a generic tool of robustness test for audio watermarking systems. It is derived from StirMark,[1] a fair benchmark for image watermarking. A number of attacks as well as attack parameters are included in StirMark for Audio v0.2, as shown in Table C.1.

Table C.1 Robustness test items in StirMark for Audio

Attack name	Description	Parameter used
AddBrumm	Add buzz or sinus tone to the sound. The unit of the three values is samples and for the frequency hertz (Hz)	AddBrummfrom AddBrummto AddBrummstep AddBrummFreq
AddDynNoise	Add a dynamic white noise part to the samples. The given parameter sets the maximum noise value	Dynnoise
AddFFTNoise	Add white noise to the samples in the FFT domain. The value "FFTNoise" sets the power of this attack to add the noise	FFTSIZE FFTNoise
AddNoise	Add white noise to the samples. The unity is in sample values. The value "0" adds nothing and "32768" the absolute distorted maximum	Noisefrom Noiseto Noisestep

(continued)

---

[1] StirMark v3.1 is a first benchmark for image watermarking released in 1999. The latest version is StirMark Benchmark 4.0, available at http://www.petitcolas.net/fabien/watermarking/stirmark/.

Y. Lin and W.H. Abdulla, *Audio Watermark: A Comprehensive Foundation Using MATLAB*,   181
DOI 10.1007/978-3-319-07974-5, © Springer International Publishing Switzerland 2015

Table C.1 (continued)

Attack name	Description	Parameter used
AddSinus	Add a sinus signal to the sound file. With this attack you can insert a disturb signal in the frequency band where the watermark is located. The unit of the frequency parameter is hertz (Hz) and samples	AddSinusFreq AddSinusAmp
Amplify	Change the loudness of audio file. For example the value "100" does not change the amplitude and a value "50" means a half loudness	Amplify
Compressor	This attack works like a compressor. You can increase or decrease the loudness of passages. The unit of the threshold is decibel (dB). The "CompressValue" describes how the sample can be changed. "2" means that the loudness of all samples in the threshold will be half. If the value is less than "1," the compressor is an expander and will increase the loudness	ThresholdDB CompressValue
CopySample	Similar to the FlippSample attack, but this attack copies the samples between the samples with a distance of FlippDist	Period FlippDist FlippCount
CutSamples	Remove samples from the audio file. If the value of "Remove" is "10000," then this attack removes every "10000" samples "RemoveNumber" samples periodically	Remove RemoveNumber
Echo	Add an echo to the sound file. The given value means the distance of the echo	Period
Exchange	Swap two sequent samples for all samples	
ExtraStereo	Increase the stereo part of the file. If the file does not have a stereo part (only mono), then this attack does not have an effect	ExtraStereofrom ExtraStereoto ExtraStereostep
FFT_HLPassQuick	Similar to the RC-HighPass and RC-LowPass attacks, but this attack is performed in the FFT domain. FFT window size can be set with the "FFTSIZE" parameter. This attack does not fade between the FFT windows, so it is possible to hear knocks	FFTSIZE HighPassFreq LowPassFreq
FFT_Invert	Invert all samples (real and imaginary part) in the FFT domain	FFTSIZE

Table C.1 (continued)

Attack name	Description	Parameter used
FFT_RealReverse	Reverse only the real part from the FFT	FFTSIZE
FFT_Stat1	Statistical attack in the FFT domain	
FFT_Test	Currently, it is to swap some samples inside from FFT	FFTSIZE
FlippSample	Swap samples inside the sound file periodically. It swaps every "Period" "FlippCount" samples with samples which have a distance of "FlippDist"	Period FlippCount FlippDist
Invert	Invert all samples in the audio file	
LSBZero	This attack sets all LSB to "0" (zero)	
Normalize	Normalize the amplitude to the maximum value	
Nothing	This attack does nothing with the audio file. The watermark should be retrieved perfectly	
RC-HighPass	Simulate a high-pass filter built with a resistance (R) and a capacitor (C)	HighPassFreq
RC-LowPass	Simulate a low-pass filter like RC-HighPass	LowPassFreq
Resampling	Change the sampling rate of sound file	SampleRate
Smooth	This attack smoothes the samples. The setting sample value depends on the samples before and after the modifying point	
Smooth2	Similar to Smooth, but the neighbor samples are valued a little bit different	
Stat1		
Stat2		
VoiceRemove	Is the opposite to ExtraStereo. This attack removes the mono part of the file (mostly where the voice is). If the file does not have a stereo part (only mono), then everything will be removed	
ZeroCross	This attack likes a limiter. If the sample value is less than the given value (threshold), all samples are set to zero	ZeroCross
ZeroLength	If a sample value is exactly "0" (zero), then this attack inserts more samples with the value "0" (zero)	ZeroLength
ZeroRemove	This attack removes all samples where the value is "0" (zero)	

Appendix D
Critical Bandwidth

See Table D.1.

Table D.1 Critical bands over the frequency spectrum [11]

Critical band rate z/Bark	Lower frequency f_l/Hz	Upper frequency f_h/Hz	Center frequency f_c/Hz	Critical bandwidth Δf/Hz
0	0	100	50	100
1	100	200	150	100
2	200	300	250	100
3	300	400	350	100
4	400	510	450	110
5	510	630	570	120
6	630	770	700	140
7	770	920	840	150
8	920	1,080	1,000	160
9	1,080	1,270	1,170	190
10	1,270	1,480	1,370	210
11	1,480	1,720	1,600	240
12	1,720	2,000	1,850	280
13	2,000	2,320	2,150	320
14	2,320	2,700	2,500	380
15	2,700	3,150	2,900	450
16	3,150	3,700	3,400	550
17	3,700	4,400	4,000	700
18	4,400	5,300	4,800	900
19	5,300	6,400	5,800	1,100
20	6,400	7,700	7,000	1,300
21	7,700	9,500	8,500	1,800
22	9,500	12,000	10,500	2,500
23	12,000	15,500	13,500	3,500
24	15,500			

Y. Lin and W.H. Abdulla, *Audio Watermark: A Comprehensive Foundation Using MATLAB*,   185
DOI 10.1007/978-3-319-07974-5, © Springer International Publishing Switzerland 2015

Appendix E
List of Audio Test Files

All the audio test samples are 44.1 kHz, 16 bit, monaural wave format files, as listed in Table E.1.

Table E.1 Descriptions of audio test files for performance evaluation

No.	Description		Durations	Waveform
A_1	Vocal	Soprano	23.6 sec	
A_2		Bass	24.9 sec	
A_3		Quartet	23.1 sec	
A_4	Percussive instrument	Hihat	2.53 sec	
A_5		Castanets	6.63 sec	
A_6		Glockenspiel1	25.9 sec	
A_7		Glockenspiel2	19.0 sec	

Y. Lin and W.H. Abdulla, *Audio Watermark: A Comprehensive Foundation Using MATLAB*, 187
DOI 10.1007/978-3-319-07974-5, © Springer International Publishing Switzerland 2015

No.	Description		Durations	Waveform
A_8	Tonal instrument	Harpsichord	16.4 sec	
A_9		Violoncello	30.1 sec	
A_{10}		Horn	25.9 sec	
A_{11}		Pipes	15.0 sec	
A_{12}		Trumpt	17.8 sec	
A_{13}		Electronic tune	35.0 sec	
A_{14}	Music	Bach	20.0 sec	
A_{15}		Pop	20.0 sec	
A_{16}		Rock	20.0 sec	
A_{17}		Jazz	20.0 sec	

Appendix F
Basic Robustness Test

Basic robustness test is applied to the watermarked audio signal S_w to inspect its capability of resisting different attacks as listed in Table F.1.

Table F.1 Descriptions of basic robustness test

Testing item	Parameters (default value)	Expression	Implementation
No attack	/	No attack	/
Noise addition	snr: signal-to-noise ratio (36 dB)	Noise (snr)	In MATLAB
Resampling	f_w: downsampling frequency (22.05 kHz)	Resampling (f_w)	Adobe Audition v3.0: 44.1 kHz $\rightarrow f_w \rightarrow$ 44.1 kHz
Requantization	Q_w: requantization bit number (8 bit)	Requantization (Q_w)	Adobe Audition v3.0: 16 bit $\rightarrow Q_w \rightarrow$ 16 bit
Amplitude scaling	A_s: rate of scaling (10 %)	Amplitude ($\pm A_s$)	Adobe Audition v3.0
Lowpass filtering	f_{cutoff}: cutoff frequency (8 kHz)	Lp filtering (f_{cutoff})	In MATLAB
DA/AD conversion	R: recording mode (line-in jack)	DA/AD (R)	Adobe Audition v3.0: play and record
Echo addition	A_m: normalized amplitude attenuation (0.3) t_d: delay time (200 ms)	Echo (A_m, t_d)	In MATLAB
Reverberation	t_{reverb}: reverberation time (1 s)	Reverb (t_{reverb})	Adobe Audition v3.0
MP3 compression	m: compression bitrate (96 kbps)	Compression I (m) Compression II (m)	Adobe Audition v3.0: .wav $\rightarrow$ .mp3 $\rightarrow$ .wav

(continued)

Y. Lin and W.H. Abdulla, *Audio Watermark: A Comprehensive Foundation Using MATLAB*,   189
DOI 10.1007/978-3-319-07974-5, © Springer International Publishing Switzerland 2015

Table F.1 (continued)

Testing item	Parameters (default value)	Expression	Implementation
Random samples cropping	n_c: no. of croppings (8) t_c: cropped interval (25 ms)	Cropping ($n_c \times t_c$)	In MATLAB
Jittering	t_j: cut clips (0.1 ms) t_f: interval of cutting (20 ms)	Jittering (t_j / t_f)	In MATLAB
Zeros inserting	n_z: no. of insertings (8) t_z: inserted interval (25 ms)	Inserting ($n_z \times t_z$)	In MATLAB
Pitch-invariant time-scale modification	P_{TSM}: percentage of time stretching ($\pm 4\%$)	PITSM (P_{TSM})	Adobe Audition v3.0
Tempo-preserved pitch-scale modification	P_{PSM}: percentage of pitch shifting ($\pm 4\%$)	TPPSM (P_{PSM})	Adobe Audition v3.0

Appendix G
Nonuniform Subbands

Audio signals used in this book are in WAVE format (44.1 kHz, 16 bit), and hence the nonuniform subbands are designed to cover $100 \sim 22,050$ Hz frequency band. When $N_{\text{subband}} = 32$, the lower/upper limits of the subbands obtained are presented in Table G.1. The number of FFT coefficients in each subband is calculated based on a frame length $N = 512$.

Table G.1 Thirty-two nonuniform subbands over the frequency spectrum

Subband index	Lower limit V^l/Hz	Upper limit V^h/Hz	Bandwidth B_w /Hz	No. of FFT coefficients
1	559.8	990.5	430.7	5
2	990.5	1,421.2	430.7	5
3	1,421.2	1,851.9	430.7	5
4	1,851.9	2,282.6	430.7	5
5	2,282.6	2,713.3	430.7	5
6	2,713.3	3,144.0	430.7	5
7	3,144.0	3,574.7	430.7	5
8	3,574.7	4,005.4	430.7	5
9	4,005.4	4,436.1	430.7	5
10	4,436.1	4,866.8	430.7	5
11	4,866.8	5,297.5	430.7	5
12	5,297.5	5,728.2	430.7	5
13	5,728.2	6,158.9	430.7	5
14	6,158.9	6,589.6	430.7	5
15	6,589.6	7,020.3	430.7	5
16	7,020.3	7,537.1	516.8	6
17	7,537.1	8,053.9	516.8	6
18	8,053.9	8,570.7	516.8	6
19	8,570.7	9,173.6	602.9	7
20	9,173.6	9,776.5	602.9	7
21	9,776.5	10,465.6	689.1	8
22	10,465.6	11,240.8	775.2	9

(continued)

Y. Lin and W.H. Abdulla, *Audio Watermark: A Comprehensive Foundation Using MATLAB*, 191
DOI 10.1007/978-3-319-07974-5, © Springer International Publishing Switzerland 2015

Table G.1 (continued)

Subband index	Lower limit V^l/Hz	Upper limit V^h/Hz	Bandwidth B_w /Hz	No. of FFT coefficients
23	11,240.8	12,016.0	775.2	9
24	12,016.0	12,791.2	775.2	9
25	12,791.2	13,738.7	947.5	11
26	13,738.7	14,686.2	947.5	11
27	14,686.2	15,633.7	947.5	11
28	15,633.7	16,753.4	1,119.7	13
29	16,753.4	17,873.1	1,119.7	13
30	17,873.1	19,165.1	1,292.0	15
31	19,165.1	20,457.1	1,292.0	15
32	20,457.1	21,835.2	1,378.1	16

References

1. H. Malik, R. Ansari, A. Khokhar, Robust audio watermarking using frequency-selective spread spectrum. IET Inform. Secur. **2**(4), 129–150 (2008)
2. S.J. Xiang, H.J. Kim, J.W. Huang, Audio watermarking robust against time-scale modification and mp3 compression. Signal Process. **88**(10), 2372–2387 (2008)
3. O.T.-C. Chen, W.-C. Wu, Highly robust, secure, and perceptual-quality echo hiding scheme. IEEE Trans. Audio Speech Lang. Process. **16**(3), 629–638 (2008)
4. X. He, *Watermarking in Audio: Key Techniques and Technologies* (Cambria Press, Youngstown, 2008)
5. W. Li, X. Y. Xue, P.Z. Lu, Localized audio watermarking technique robust against time-scale modification. IEEE Trans. Multimed. **8**(1), 60–69 (2006)
6. M.F. Mansour, A.H. Tewfik, Data embedding in audio using time-scale modification. IEEE Trans. Speech Audio Process. **13**(3), 432–440 (2005)
7. N. Cvejic, T. Seppanen, Robust audio watermarking in wavelet domain using frequency hopping and patchwork method, in *Proceedings of the 3rd International Symposium on Image and Signal Processing and Analysis*, 2003, pp. 251–255
8. D. Kirovsk, H.S. Malvar, Spread-spectrum watermarking of audio signals. IEEE Trans. Signal Process. **51**(4), 1020–1033 (2003)
9. R. Tachibana, S. Shimizu, S. Kobayashi, An audio watermarking method using a two-dimensional pseudo-random array. Signal Process. **82**(10), 1455–1469 (2002)
10. R. Tachibana, Improving audio watermarking robustness using stretched patterns against geometric distortion, in *Proceedings of IEEE Pacific-Rim Conference on Multimedia (PCM)*, 2002, pp. 647–654
11. E. Zwicker, H. Fastl, *Psychoacoustics: Facts and Models* (Springer, Berlin, 1990)
12. G. Widmer, D. Rocchesso, V. Välimäki, C. Erkut, F. Gouyon, D. Pressnitzer, et al., Sound and music computing: research trends and some key issues. J New Music Res. **36**, 169–184 (2007)
13. I.J. Cox, M.L. Miller, J.A. Bloom, J. Fridrich, T. Kalker, *Digital Watermarking and Steganography* (Morgan Kaufmann Publishers, San Francisco, 2008)
14. S. Katzenbeisser, F.A.P. Petitcolas (eds.), *Information Hiding Techniques for Steganography and Digital Watermarking* (Artech House, Boston, 2000)
15. N.F. Johnson, Z. Duric, S. Jajodia, *Information Hiding: Steganography and Watermarking - Attacks and Countermeasures* (Kluwer Academic, Boston, 2001)
16. R. Walker, Audio watermarking. Technical Report, BBC R&D (2004) [Online], http://www.bbc.co.uk/rd/pubs/whp/whp-pdf-files/WHP057.pdf
17. S.P. Mohanty, Digital watermarking: a tutorial review. Technical Report, University of South Florida (1999) [Online], http://www.cs.unt.edu/smoh-anty/research/Reports/MohantyWatermarkingSurvey1999.pdf

Y. Lin and W.H. Abdulla, *Audio Watermark: A Comprehensive Foundation Using MATLAB*, 193
DOI 10.1007/978-3-319-07974-5, © Springer International Publishing Switzerland 2015

18. M.D. Swanson, M. Kobayashi, A.H. Tewfik, Multimedia data: embedding and watermarking technologies. Proc. IEEE **86**(6), 1064–1087 (1998)
19. Y.Q. Lin, W.H. Abdulla, Audio watermarking for copyrights protection. Technical Report SoE-650, School of Engineering, The University of Auckland (2007)
20. F.A.P. Petitcolas, R.J. Anderson, M.G. Kuhn, Information hiding: a survey. Proc. IEEE **87**(7), 1062–1078 (1999)
21. S.A. Craver, M. Wu, B. Liu, What can we reasonably expect from watermarks? in *Proceedings of IEEE Workshop on Applications of Signal Processing to Audio and Acoustics*, 2001, pp. 223–226
22. L.d.C.T. Gomes, P. Cano, E. Gómez, M. Bonnet, E. Batlle, Audio watermarking and fingerprinting: for which applications? J. New Music Res. **32**(1), 65–81 (2003)
23. F. Kurth, M. Muller, Efficient index-based audio matching. IEEE Trans. Audio Speech Lang. Process. **16**(2), 382–395 (2008)
24. M. Barni, F. Bartolini, *Watermarking Systems Engineering: Enabling Digital Assets Security and Other Applications* (Marcel Dekker, New York, 2004)
25. J.-S. Pan, H.-C. Huang, L.C. Jain (eds.), *Intelligent Watermarking Techniques* (World Scientific, River Edge, 2004)
26. J. Seitz (ed.), *Digital Watermarking for Digital Media* (Information Science Publishers, Hershey, 2005)
27. B. Furht, D. Kirovski (eds.), *Multimedia Watermarking Techniques and Applications* (Auerbach Publications, Boca Raton, 2006)
28. F. Hartung, M. Kutter, Multimedia watermarking techniques. Proc. IEEE **87**(7), 1079–1107 (1999)
29. T. Page, Digital watermarking as a form of copyright protection. Comput. Law Secur. Rep. **14**(6), 390–392 (1998)
30. N. Cvejic, T. Seppanen (eds.), *Digital Audio Watermarking Techniques and Technologies: Applications and Benchmarks* (Information Science Reference, Hershey, 2008)
31. M. Arnold, M. Schmucker, S.D. Wolthusen, *Techniques and Applications of Digital Watermarking and Content Protection* (Artech House, Boston, 2003)
32. L. Boney, A.H. Tewfik, K.N. Hamdy, Digital watermarks for audio signals, in *Proceedings of IEEE International Conference on Multimedia Computing and Systems*, 1996, pp. 473–480
33. C.-P. Wu, P.-C. Su, C.-C.J. Kuo, Robust and efficient digital audio watermarking using audio content analysis, in *Proceedings of SPIE Security and Watermarking of Multimedia Contents II*, vol. 3971, 2000 , pp. 382–392
34. W.-N. Lie, L.-C. Chang, Robust and high-quality time-domain audio watermarking subject to psychoacoustic masking, in *Proceedings of IEEE International Symposium on Circuits and Systems (ISCAS)*, 2001, pp. 45–48
35. S.J. Xiang, J.W. Huang, Histogram-based audio watermarking against time-scale modification and cropping attacks. IEEE Trans. Multimed. **9**(7), 1357–1372 (2007)
36. W. Bender, D. Gruhl, N. Morimoto, A. Lu, Techniques for data hiding. IBM Syst. J. **35**(3 & 4), 313–336 (1996)
37. F.A.P. Petitcolas, Watermarking schemes evaluation. IEEE Signal Process. Mag. **17**(5), 58–64 (2000)
38. A. Lang, J. Dittmann, Transparency and complexity benchmarking of audio watermarking algorithms issues, in *Proceedings of Workshop on Multimedia and Security*, 2006, pp. 190–201
39. M. Arnold, Subjective and objective quality evaluation of watermarked audio tracks, in *Proceedings of International Conference on Web Delivering of Music (WEDELMUSIC)*, 2002, pp. 161–167
40. Announcement of Evaluation Test Results for "STEP 2000". JASRAC and NRI (2000) [Online], http://www.jasrac.or.jp/watermark/ehoukoku.htm
41. A. Garay Acevedo, Audio watermarking quality evaluation, in *e-Business and Telecommunication Networks*, ed. by J. Ascenso et al. (Springer, Netherlands, 2006), pp. 272–283

42. G. Stoll, F. Kozamernik, EBU listening tests on internet audio codecs. *EBU Technical Review*, 2000
43. ITU-R Recommendation BS.1116-1, *ITU-R Recommendation BS.1116-1: Methods for the Subjective Assessment of Small Impairments in Audio Systems Including Multichannel Sound Systems*, 1997
44. ITU-R Recommendation BS.1534-1, *ITU-R Recommendation BS.1534-1: Method for the Subjective Assessment of Intermediate Quality Level of Coding Systems*, 2003
45. ITU-R Recommendation BS.1284-1, *ITU-R Recommendation BS.1284-1: General methods for the subjective assessment of sound quality*, 2003
46. J.G. Beerends, Audio quality determination based on perceptual measurement techniques, in *Applications of Digital Signal Processing to Audio and Acoustics*, ed. by M. Kahrs, K. Brandenburg (Kluwer Academic, Boston, 1998), pp. 1–38
47. A. Lerch, Software: EAQUAL - Evaluation of Audio Quality, v.0.1.3alpha ed. (2002) [Online], http://www.rarewares.org/others.php
48. P. Kabal, An examination and interpretation of ITU-R BS.1387: Perceptual evaluation of audio quality. Technical Report, TSP Lab, McGill University (2003) [Online], http://www-mmsp.ece.mcgill.ca/Documents
49. R. Huber, B. Kollmeier, PEMO-Q: a new method for objective audio quality assessment using a model of auditory perception. IEEE Trans. Audio Speech Lang. Process. **14**(6), 1902–1911 (2006) [Online]. http://www.hoertech.de/web-en/produkte/downloads.shtml
50. S.R. Quackenbush, T.P. Barnwell III, M.A. Clements, *Objective Measures of Speech Quality* (Prentice Hall, Englewood Cliffs, 1988)
51. M. Bosi, R.E. Goldberg, *Introduction to Digital Audio Coding and Standards* (Kluwer Academic, Boston, 2003)
52. W.J. Vincoli (ed.), *Lewis' Dictionary of Occupational and Environmental Safety and Health* (Lewis Publishers, Boca Raton, 2000)
53. K. Johnson, *Acoustic and Auditory Phonetics* (Blackwell Publisher, Malden, 2003)
54. P.H. Lindsay, D.A. Norman, *Human Information Processing: An Introduction to Psychology* (Academic, New York, 1977)
55. T.S. Gunawan, Audio compression and speech enhancement using temporal masking models. Ph.D. dissertation, The University of New South Wales, 2007
56. [Online]. Available: http://projects.cbe.ab.ca/Diefenbaker/Biology/Bio%20Website%20Final/notes/nervous_system/Image59.gif
57. M.W. Levine, *Levine and Shefner's Fundamentals of Sensation and Perception* (Oxford University Press, Oxford, 2000)
58. W.A. Yost, D.W. Nielsen, *Fundamentals of Hearing: An Introduction* (Holt, Rinehart and Winston, New York, 1977)
59. E.A.G. Shaw, Earcanal pressure generated by a free sound field. J. Acoust. Soc. Am. **39**(3), 465–470 (1966)
60. B.C.J. Moore, *An Introduction to the Psychology of Hearing* (Academic, New York, 2003)
61. [Online]. Available: http://www.chicagoear.com/images/earworks.gif
62. T.D. Rossing (ed.), *Handbook of Acoustics* (Springer, Heidelberg, 2007)
63. [Online]. Available: http://www2.ph.ed.ac.uk/AardvarkDeployments/Public/67158/views/workspace/dwatts1/66265/inner.node/les/MusicalAcoustics/CourseNotes/PropertiesoftheEar/web.html
64. [Online]. Available: http://www.ai.rug.nl/acg/cpsp/docs/cochleaModel.html
65. I.J. Hirsh, *The Measurement of Hearing* (McGraw-Hill, New York, 1952)
66. H. Fletcher, W.A. Munson, Loudness, its definition, measurement and calculation. J. Acoust. Soc. Am. **5**(2), 82–108 (1933)
67. Y.H. Kim, H.I. Kang, K.I. Kim, S.-S. Han, A digital audio watermarking using two masking effects, in *Advances in Multimedia Information Processing - PCM 2002*, ed. by Y.-C. Chen, L.-W. Chang, H.C.-T. Lecture Notes in Computer Science, vol. 2532 (Springer, Berlin/Heidelberg, 2002), pp. 105–115

68. X.M. Quan, H.B. Zhang, Statistical audio watermarking algorithm based on perceptual analysis, in *Proceedings of the 5th ACM Workshop on Digital Rights Management*, 2005, pp. 112–118

69. E. Ambikairajah, A.G. Davis, W.T.K. Wong, Auditory masking and MPEG-1 audio compression. Electron. Comm. Eng. J. **9**, 165–173 (1997)

70. A. Spanias, T. Painter, V. Atti, *Audio Signal Processing and Coding* (Wiley-Interscience, Hoboken, 2007)

71. M.D. Swanson, B. Zhu, A.H. Tewfik, L. Boney, Robust audio watermarking using perceptual masking. Signal Process. **66**(3), 337–355 (1998)

72. R.A. Garcia, Digital watermarking of audio signals using a psychoacoustic auditory model and spread spectrum theory. *AES E-Library*, 1999

73. S. Ratanasanya, S. Poomdaeng, S. Tachphetpiboon, T. Amornraksa, New psychoacoustic models for wavelet based audio watermarking, in *IEEE International Symposium on Communications and Information Technology (ISCIT)*, vol. 1, pp. 602–605, 2005

74. ISO/IEC IS 11172-3, *Information Technology - Coding of Moving Picture and Associated Audio for Digital Storage Media Up To About 1.5Mbit/s, Part 3: Audio* (BSI, London, 1993)

75. K.C. Pohlmann, *Principles of Digital Audio* (McGraw-Hill, New York, 2000)

76. D. Pan, A tutorial on MPEG/audio compression. IEEE Multimed. **2**, 60–74 (1995)

77. F.A.P. Petitcolas, MPEG for Matlab, v.1.2.8 ed. (2003) [Online], http://www.petitcolas.net/fabien/software/mpeg

78. C.-Y. Lin, An investigation into perceptual audio coding and the use of auditory gammatone filterbanks. Master's thesis, The University of Auckland, 2007

79. SQAM - Sound Quality Assessment Material, European Broadcasting Union (EBU) [Online], http://sound.media.mit.edu/mpeg4/audio/sqam

80. A. Takahashi, R. Nishimura, Y. Suzuki, Multiple watermarks for stereo audio signals using phase-modulation techniques. IEEE Trans. Signal Process. **53**(2), 806–815 (2005)

81. P. Liew, M. Armand, Inaudible watermarking via phase manipulation of random frequencies. Multimed. Tools Appl. **35**(3), 357–377 (2007)

82. A. Piva, M. Barni, F. Bartolini, A. De Rosa, Data hiding technologies for digital radiography. IEE Proc. Vision Image Signal Process. **152**(5), 604–610 (2005)

83. B. Chen, G.W. Wornell, Quantization index modulation: a class of provably good methods for digital watermarking and information embedding. IEEE Trans. Inform. Theory **47**(4), 1423–1443 (2001)

84. A. Zaidi, R. Boyer, P. Duhamel, Audio watermarking under desynchronization and additive noise attacks. IEEE Trans. Signal Process. **54**(2), 570–584 (2006)

85. D. Lam, Audio watermarking. COMPSYS401A Project, The University of Auckland, 2003

86. S. Saito, T. Furukawa, K. Konishi, A digital watermarking for audio data using band division based on QMF bank, in *Proceedings of IEEE International Conference on Acoustics, Speech and Signal Processing (ICASSP)*, vol. 4, 2002, pp. 3473–3476

87. A.V. Oppenheim, R.W. Schafer, *Discrete-Time Signal Processing* (Prentice Hall, Englewood Cliffs, 1989)

88. S.-S. Kuo, J.D. Johnston, W. Turin, S.R. Quackenbush, Covert audio watermarking using perceptually tuned signal independent multiband phase modulation. Proc. ICASSP **2**, 1753–1756 (2002)

89. I.J. Cox, J. Kilian, F.T. Leighton, T. Shamoon, Secure spread spectrum watermarking for multimedia. IEEE Trans. Image Process. **6**(12), 1673–1687 (1997)

90. H.J. Kim, Audio watermarking techniques, in *Proceedings of Pacific Rim Workshop on Digital Steganography*, 2003

91. H. Malik, A. Khokhar, A. Rashid, Robust audio watermarking using frequency selective spread spectrum theory. Proc. ICASSP **5**, 385–388 (2004)

92. N. Cvejic, T. Seppanen, Spread spectrum audio watermarking using frequency hopping and attack characterization. Signal Process. **84**(1), 207–213 (2004)

93. J. Seok, J. Hong, J. Kim, A novel audio watermarking algorithm for copyright protection of digital audio. ETRI J. **24**(3), 181–189 (2002)

94. L.R. Rabiner, R.W. Schafer, *Digital Processing of Speech Signals* (Prentice Hall, Englewood Cliffs, 1978)

95. X. Li, H.H. Yu, Transparent and robust audio data hiding in cepstrum domain, in *Proceedings of IEEE International Conference on Multimedia and Expo (ICME)*, vol. 1, 2000, pp. 397–400

96. S.-K. Lee, Y.-S. Ho, Digital audio watermarking in the cepstrum domain. IEEE Trans. Consumer Electron. **46**(3), 744–750 (2000)

97. C.-T. Hsieh, P.-Y. Sou, Blind cepstrum domain audio watermarking based on time energy features, in *Proceedings of International Conference on Digital Signal Processing (DSP)*, vol. 2, 2002, pp. 705–708

98. L.L. Cui, S.X. Wang, T.F. Sun, The application of binary image in digital audio watermarking, in *Proceedings of International Conference on Neural Networks and Signal Processing*, vol. 2, 2003, pp. 1497–1500

99. K. Gopalan, Audio steganography by cepstrum modification. Proc. ICASSP **5**, 481–484 (2005)

100. K. K. Parhi, T. Nishitani, *Digital Signal Processing for Multimedia Systems* (CRC Press, New York, 1999)

101. W.Y. Hwang, H.I. Kang, S.S. Han, K.I. Kim, H.S. Kang, Robust audio watermarking using both DWT and masking effect, in *Digital Watermarking, LNCS 2939*, ed. by T. Kalker et al. (Springer, Berlin/Heidelberg, 2004), pp. 382–389

102. A. Prochazka, J. Uhlir, P.W.J. Rayner, N.G. Kingsbury, *Signal Analysis and Prediction* (Birkhäuser, Boston, 1998)

103. X. He, M.S. Scordilis, An enhanced psychoacoustic model based on the discrete wavelet packet transform. J. Franklin Inst. **343**(7), 738–755 (2006)

104. C.-S. Ko, K.-Y. Kim, R.-W. Hwang, Y.-S. Kim, S.-B. Rhee, Robust audio watermarking in wavelet domain using pseudorandom sequences, in *Proceedings of Annual International Conference on Computer and Information Science (ACIS)*, 2005, pp. 397–401

105. P. Artameeyanant, Wavelet audio watermark robust against MPEG compression, in *SICE Annual Conference*, pp. 1414–1417, 2007

106. H.O. Kim, B.K. Lee, N. Lee, Wavelet-based audio watermarking techniques: robustness and fast synchronization [Online], http://amath.kaist.ac.kr/research/paper/01-11.pdf

107. W. Li, X.Y. Xue, An audio watermarking technique that is robust against random cropping. Comput. Music J. **27**(4), 58–68 (2003)

108. H.O. Oh, J.W. Seok, J.W. Hong, D.H. Youn, New echo embedding technique for robust and imperceptible audio watermarking. Proc. ICASSP **3**, 1341–1344 (2001)

109. D. Gruhl, A. Lu, W. Bender, Echo hiding, in *Information Hiding*, ed.b y R. Anderson. Lecture Notes in Computer Science, vol. 1174 (Springer, Berlin/Heidelberg, 1996), pp. 295–315

110. H.J. Kim,Y.H. Choi, A novel echo-hiding scheme with backward and forward kernels. IEEE Trans. Circ. Syst. Video Tech. **13**(8), 885–889 (2003)

111. B.-S. Ko, R. Nishimura, Y. Suzuki, Time-spread echo method for digital audio watermarking. IEEE Trans. Multimed. **7**(2), 212–221 (2005)

112. B.-S. Ko, R. Nishimura, Y. Suzuki, Log-scaling watermark detection in digital audio watermarking. Proc. ICASSP **3**, 81–84 (2004)

113. D. Coltuc, P. Bolon, Robust watermarking by histogram specification, in *Proceedings of International Conference on Image Processing (ICIP)*, vol. 2, 1999, pp. 236–239

114. M. Mese, P.P. Vaidyanathan, Optimal histogram modification with MSE metric. Proc. ICASSP **3**, 1665–1668 (2001)

115. E. Chrysochos, V. Fotopoulos, A.N. Skodras, M. Xenos, Reversible image watermarking based on histogram modification, in *Proceedings of the 11th Panhellenic Conference on Informatics (PCI)*, vol. B, 2007, pp. 93–104

116. G.R. Xuan, Q.M. Yao, C.Y. Yang, J.J. Gao, P.Q. Chai, Y. Shi, Z.C. Ni, Lossless data hiding using histogram shifting method based on integer wavelets, in *Digital Watermarking*, ed. by Y.Q. Shi, B. Jeon. Lecture Notes in Computer Science, vol. 4283 (Springer, Berlin/Heidelberg, 2006), pp. 323–332

117. S.J. Xiang, J.W. Huang, R. Yang, Time-scale invariant audio watermarking based on the statistical features in time domain, in *Information Hiding*, ed. by J. Camenisch et al. Lecture Notes in Computer Science, vol. 4437 (Springer, Berlin/Heidelberg, , 2007), pp. 93–108. Matlab implementation available at http://cist.korea.ac.kr/xiangshijun/

118. D.R. Smith, *Digital Transmission Systems* (Kluwer Academic, Boston, 2004)

119. H. Farid, Detecting hidden messages using higher-order statistical models. Proc. ICIP **2**, 905–908 (2002)

120. M. Alghoniemy, A.H. Tewfik, Image watermarking by moment invariants. Proc. ICIP **2**, 73–76 (2000)

121. S.J. Xiang, J.W. Huang, R. Yang, C.T. Wang, H.M. Liu, Robust audio watermarking based on low-order zernike moments, in *Digital Watermarking*, ed. by Y.Q. Shi, B. Jeon. Lecture Notes in Computer Science, vol. 4283 (Springer, Berlin/Heidelberg, 2006), pp. 226–240

122. P. Bas, J.-M. Chassery, B. Macq, Geometrically invariant watermarking using feature points. IEEE Trans. Image Process. **11**(9), 1014–1028 (2002)

123. F.-S. Wei, F. Xue, M.Y. Li, A blind audio watermarking scheme using peak point extraction. Proc. ISCAS **5**, 4409–4412 (2005)

124. W.H. Abdulla, Auditory based feature vectors for speech recognition systems, in *Advances in Communications and Software Technologies*, ed. by N.E. Mastorakis, V.V. Kluev (WSEAS Press, Greece, 2002), pp. 231–236

125. Y.Q. Lin, W.H. Abdulla, Robust audio watermarking technique based on Gammatone filterbank and coded-image, in *Proceedings of International Symposium on Signal Processing and Its Applications (ISSPA)*, 2007

126. D. Bailey, W. Cammack, J. Guajardo, C. Paar, Cryptography in modern communication systems, in *TI DSPS FEST*, pp. 1–15, 1999

127. Y.Q. Lin, W.H. Abdulla, A secure and robust audio watermarking scheme using multiple scrambling and adaptive synchronization, in *Proceedings of the 6th International Conference on Information, Communications and Signal Processing (ICICS)*, 2007

128. Y.Q. Lin, W.H. Abdulla, Y. Ma, Audio watermarking detection resistant to time and pitch scale modification, in *Proceedings of IEEE International Conference on Signal Processing and Communications (ICSPC)*, 2007, pp. 1379–1382

129. M. Kahrs, K. Brandenburg, *Applications of Digital Signal Processing to Audio and Acoustics* (Kluwer Academic, Boston, 1998)

130. C.-W. Tang, H.-M. Hang, A feature-based robust digital image watermarking scheme. IEEE Trans. Signal Process. **51**(4), 950–959 (2003)

131. Y.Q. Lin, W.H. Abdulla, Multiple scrambling and adaptive synchronization for audio watermarking, in *Digital Watermarking*, ed. by Y.Q. Shi, H.-J. Kim, S. Katzenbeisser. Lecture Notes in Computer Science, vol. 5041 (Springer, Berlin/Heidelberg, 2007), pp. 440–453

132. T. Acharya, A.K. Ray, *Image Processing: Principles and Applications* (Wiley, Hoboken, 2005)

133. N. Collins, *Introduction to Computer Music* (Wiley, New York, 2009)

134. A. Lang, Documentation for Stirmark for Audio (2002) [Online], http://amsl-smb.cs.uni-magdeburg.de/stirmark/doc/index.html

135. H. Zhao, M. Wu, Z.J. Wang, K.J.R. Liu, Nonlinear collusion attacks on independent fingerprints for multimedia. Proc. ICASSP **5**, 664–667 (2003)

136. J.H.L. Hansen, B.L. Pellom, An effective quality evaluation protocol for speech enhancement algorithms, in *Proceedings of International Conference on Spoken Language Processing (INTERSPEECH)*, vol. 7, 1998, pp. 2819–2822

137. F. Mustiere, M. Bouchard, M. Bolic, Quality assessment of speech enhanced using particle filters. Proc. ICASSP **3**, 1197–1200 (2007)

138. Y. Hu, P.C. Loizou, Evaluation of objective quality measures for speech enhancement. IEEE Trans. Audio Speech Lang. Process. **16**(1), 229–238 (2008)

139. W.M. Liu, K.A. Jellyman, J.S.D. Mason, N.W.D. Evans, Assessment of objective quality measures for speech intelligibility estimation. Proc. ICASSP **1**, 1225–1228 (2006)

140. L. Di Persia, M. Yanagida, H.L. Rufiner, D. Milone, Objective quality evaluation in blind source separation for speech recognition in a real room. Signal Process. **87**(8), 1951–1965 (2007)
141. L. Di Persia, D. Milone, H.L. Rufiner, M. Yanagida, Perceptual evaluation of blind source separation for robust speech recognition. Signal Process. **88**(10), 2578–2583 (2008)
142. Y.Q. Lin, W.H. Abdulla, Perceptual evaluation of audio watermarking using objective quality measures, in *Proceedings of ICASSP*, 2008, pp. 1745–1748
143. Y. Lin, W. Abdulla, Objective quality measures for perceptual evaluation in digital audio watermarking. IET - Signal Process. **5**(7), 623–631 (2011)
144. T. Rohdenburg, V. Hohmann, B. Kollmeier, Objective perceptual quality measures for the evaluation of noise reduction schemes, in *Proceedings of the 9th International Workshop on Acoustic Echo and Noise Control (IWAENC)*, 2005, pp. 169–172
145. SDMI Portable Device Specification, Part 1 (Version 1.0). SDMI (1999) [Online]. http://ntrg.cs.tcd.ie/undergrad/4ba2.01/group10/technology.html
146. Call for Proposals for Phase II Screening Technology (Version 1.0). SDMI (2000) [Online]. http://ntrg.cs.tcd.ie/undergrad/4ba2.01/group10/technology.html